TRANSPORTATION ISSUES, POLICIES AND R&D

VEHICLE-TO-INFRASTRUCTURE TECHNOLOGIES

ELEMENTS, DEPLOYMENT CHALLENGES, AND SAFETY APPLICATIONS

TRANSPORTATION ISSUES, POLICIES AND R&D

Additional books in this series can be found on Nova's website under the Series tab.

Additional e-books in this series can be found on Nova's website under the eBooks tab.

TRANSPORTATION ISSUES, POLICIES AND R&D

VEHICLE-TO-INFRASTRUCTURE TECHNOLOGIES

ELEMENTS, DEPLOYMENT CHALLENGES, AND SAFETY APPLICATIONS

ALBERT BUCHANAN
EDITOR

nova publishers
New York

Library of Congress Cataloging-in-Publication Data

ISBN: 978-1-63485-111-4

Published by Nova Science Publishers, Inc. † New York

CONTENTS

PREFACE

Over the past two decades, automobile crash-related fatality and injury rates have declined over 34 and 40 percent respectively, due in part to improvements in automobile safety. To further improve traffic safety and provide other transportation benefits, the Department of Transportation (DOT) is promoting the development of vehicle-to-infrastructure (V2I) technologies. Among other things, V2I technologies would allow roadside devices and vehicles to communicate and alert drivers of potential safety issues, such as if they are about to run a red light. GAO was asked to review V2I deployment. This book addresses the status of V2I technologies; challenges that could affect the deployment of V2I technologies, and DOT efforts to address these challenges; and what is known about the potential benefits and costs of V2I technologies.

In: Vehicle-to-Infrastructure Technologies ISBN: 978-1-63485-111-4
Editor: Albert Buchanan

Chapter 1

INTELLIGENT TRANSPORTATION SYSTEMS: VEHICLE-TO-INFRASTRUCTURE TECHNOLOGIES EXPECTED TO OFFER BENEFITS, BUT DEPLOYMENT CHALLENGES EXIST*

United States Government Accountability Office

WHY GAO DID THIS STUDY

Over the past two decades, automobile crash-related fatality and injury rates have declined over 34 and 40 percent respectively, due in part to improvements in automobile safety. To further improve traffic safety and provide other transportation benefits, DOT is promoting the development of V2I technologies. Among other things, V2I technologies would allow roadside devices and vehicles to communicate and alert drivers of potential safety issues, such as if they are about to run a red light. GAO was asked to review V2I deployment.

This report addresses: (1) the status of V2I technologies; (2) challenges that could affect the deployment of V2I technologies, and DOT efforts to

* This is an edited, reformatted and augmented version of a United States Government Accountability Office, Publication No. GAO-15-775, dated September 2015.

address these challenges; and (3) what is known about the potential benefits and costs of V2I technologies.

GAO reviewed documentation on V2I from DOT, automobile manufacturers, industry associations, and state and local agencies. In addition, GAO interviewed DOT, Federal Communication Commission (FCC), and National Telecommunications Information Administration (NTIA) officials. GAO also conducted structured interviews with 21 experts from a variety of subject areas related to V2I. The experts were chosen based on recommendations from the National Academy of Sciences and other factors.

DOT, NTIA, and the FCC reviewed a draft of this report. DOT and NTIA provided technical comments, which were incorporated as appropriate. FCC did not provide comments.

WHAT GAO FOUND

Vehicle-to-infrastructure (V2I) technologies allow roadside devices to communicate with vehicles and warn drivers of safety issues; however, these technologies are still developing. According to the Department of Transportation (DOT), extensive deployment may occur over the next few decades. DOT, state, and local-transportation agencies; researchers; and private-sector stakeholders are developing and testing V2I technologies through test beds and pilot deployments. Over the next 5 years, DOT plans to provide up to $100 million through its Connected Vehicle pilot program for projects that will deploy V2I technologies in real-world settings. DOT and other stakeholders have also provided guidance to help state and local agencies pursue V2I deployments, since it will be up to these agencies to voluntarily deploy V2I technologies.

According to experts and industry stakeholders GAO interviewed, there are a variety of challenges that may affect the deployment of V2I technologies including: (1) ensuring that possible sharing with other wireless users of the radio-frequency spectrum used by V2I communications will not adversely affect V2I technologies' performance; (2) addressing states and local agencies' lack of resources to deploy and maintain V2I technologies; (3) developing technical standards to ensure interoperability; (4) developing and managing data security and addressing public perceptions related to privacy; (5) ensuring that drivers respond appropriately to V2I warnings; and (6) addressing the uncertainties related to potential liability issues posed by V2I. DOT is collaborating with the automotive industry and state transportation officials, among others, to identify potential solutions to these challenges.

The full extent of V2I technologies' benefits and costs is unclear because test deployments have been limited thus far; however, DOT has supported initial research into the potential benefits and costs. Experts GAO spoke to and research GAO reviewed indicate that V2I technologies could provide safety, mobility, environmental, and operational benefits, for example by: (1) alerting drivers to potential dangers, (2) allowing agencies to monitor and address congestion, and (3) providing driving and route advice. V2I costs will include the initial non-recurring costs to deploy the infrastructure and the recurring costs to operate and maintain the infrastructure. While some organizations have estimated the potential average costs for V2I deployments, actual costs will depend on a variety of factors, including where the technology is installed, and how much additional infrastructure is needed to support the V2I equipment.

Source: GAO analysis of Department of Transportation documents. | GAO-15-775.

Example of a Vehicle-to-Infrastructure Application.

ABBREVIATIONS

AASHTO	American Association of State Highway and Transportation Officials
ANPRM	Advanced Notice of Proposed Rulemaking
CAMP	Crash Avoidance Metrics Partners, LLC
CO-Pilot	Cost Overview for Planning Ideas and Logical Organization Tool
DOT	U.S. Department of Transportation
DSRC	dedicated short-range communications
GHz	gigahertz
FCC	Federal Communications Commission
FHWA	Federal Highway Administration
IEEE	Institute of Electrical and Electronics Engineers

ITS	intelligent transportation system
ITS	America Intelligent Transportation Society of America
ITS-JPO	Intelligent Transportation Systems-Joint Program Office
MHz	megahertz
MLIT	Ministry of Land, Infrastructure, Transport and Tourism (Japan)
NCHRP	National Cooperative Highway Research Program
NHTSA	National Highway Traffic Safety Administration
NTIA	National Telecommunications and Information Administration
RSE	roadside equipment
RSU	roadside unit
SAE	Society for Automotive Engineers International
SCMS	Security Credential Management System
SPaT	Signal Phase and Timing
TMC	traffic management center
V2I	vehicle-to-infrastructure
V2V	vehicle-to-vehicle
VICS	Vehicle Information and Communication System (Japan)

* * *

September 15, 2015

The Honorable Barbara Comstock
Chairwoman
The Honorable Daniel Lipinski
Ranking Member
Subcommittee on Research and Technology Committee on Science, Space, and Technology
House of Representatives

The Honorable Larry Bucshon
House of Representatives

Over the past two decades, automobile crash-related fatality and injury rates have declined nearly 34 percent and 40 percent, respectively, due in part to automobile safety features like safety belts and airbags.[1] The U.S. Department of Transportation (DOT) is working to further improve traffic safety through its connected vehicle research program, which aims to develop innovative technologies that enable vehicles, road infrastructure, and personal communications devices to wirelessly communicate and warn drivers and pedestrians of potential accidents. For example, DOT is collaborating with the automobile industry, academic institutions, technology firms, and state and local agencies to develop vehicle-to-infrastructure (V2I) technologies that allow vehicles to "communicate" with road infrastructure (such as traffic signals) through the wireless exchange of data. These technologies can enable the development of V2I software applications[2] that could, among other things: warn drivers of upcoming road conditions, such as work zones, or that they are approaching a curve at an unsafe speed; adjust traffic signal lights to provide priority to emergency vehicles or to address congestion; advise drivers about upcoming traffic and alternative routes; and provide driving advice to minimize stop-and-go driving.[3] For example, in 2011, Japan implemented V2I through the deployment of the ITS Spot system. ITS Spot uses roadside equipment to collect and share data with vehicles in order to provide three basic services to drivers: dynamic route guidance, safe driving support, and electronic toll collection. Japan's extensive V2I network includes roughly 55,000 pieces of V2I equipment on local roads and 1,600 pieces of V2I equipment on its approximately 11,000 kilometers of expressways. Similarly, the Netherlands, Germany, and Austria are working to develop a European smart corridor that will provide drivers information on road work and upcoming traffic, among other things. Since V2I technologies are still in development in the United States and rely on the exchange of information between vehicles and infrastructure, developing and deploying V2I will require the collaboration of a number of stakeholders, particularly state and local agencies, as well as auto manufacturers.

In light of research showing the potential for V2I technologies to reduce traffic accidents and fatalities, as well as questions raised regarding potential technological and policy challenges, you asked us to review issues related to V2I technologies. We examined: (1) the status of V2I technologies; (2) the challenges, if any, that could affect the deployment of V2I technologies, and DOT efforts to address these challenges; and (3) what is known about the potential benefits and costs of V2I technologies. To address these issues, we reviewed documentation relevant to the V2I technology research efforts of DOT, state and local governments, and the automobile industry, including DOT's *2015 Federal Highway Administration (FHWA) V2I Draft Deployment Guidance and Products*[4] and the American Association of State Highway and Transportation Officials' (AASHTO) *National Connected Vehicle Field*

Infrastructure Footprint Analysis. We interviewed officials from DOT's Office of the Assistant Secretary for Research and Technology, Intelligent Transportation Systems-Joint Program Office (ITS-JPO), FHWA, National Highway Traffic Safety Administration (NHTSA), and the Volpe National Transportation Systems Center about these efforts. In addition to DOT and its agencies, we also interviewed an additional 12 stakeholders that were involved in V2I efforts, such as associations representing state transportation agencies and engineers.[5] We interviewed officials at all seven V2I test beds located in Virginia, Michigan, Florida, Arizona, California, and New York.[6] We conducted site visits to three of the seven test beds—the Safety Pilot in Ann Arbor, Michigan, and the test beds in Southeast Michigan and Northern Virginia. We selected the three site visit locations based on which had the most advanced technology according to DOT and state officials. We used our interviews with stakeholders to help us understand the issues, and developed a structured set of questions for interviews with 21 experts, nine of whom were identified by the National Academy of Sciences. We selected an additional 12 experts based on the following factors: (1) their personal involvement in the deployment of V2I technologies; (2) recommendations from federal agencies and industry associations; and, (3) experts' involvement in a professional affiliation such as a V2I consortium or group dedicated to these technologies, or expertise on a specific challenge affecting V2I (e.g., privacy). The 21 experts we selected included domestic automobile manufacturers, V2I equipment suppliers, state and local government officials, privacy experts, global industry organizations responsible for developing technology standards, and academic researchers with relevant expertise. During these interviews we asked, among other things, for experts' views on the state of development and deployment of V2I technologies (including DOT's role in this process), the potential benefits of V2I technologies, and their potential costs.[7] In our report, we use the term "experts" to refer to the 21 selected individuals we interviewed using a structured set of questions; we use the term "stakeholders" to refer to those individuals we spoke with, but that were not interviewed using the structured set of questions. The viewpoints gathered through our expert interviews represent the viewpoints of these specific individuals and cannot be generalized to a broader population.

We also interviewed officials from the Federal Communications Commission (FCC) and National Telecommunications and Information Administration (NTIA) within the Department of Commerce regarding challenges related to the potential for spectrum sharing with V2I technologies. Finally, we conducted a site visit to Japan because of its years of experience with deployment and maintenance of its national V2I system.[8] During our site visit, we interviewed Japanese government officials responsible for V2I and auto manufacturers on topics similar to those discussed with U.S. experts, including V2I deployment efforts, benefits, costs, and challenges. Information about Japan's V2I efforts provides an illustrative example from which to draw information on the potential benefits, costs, and challenges of deploying V2I technologies in the United States. Further details about our scope and methodology can be found in appendix I.

We conducted this performance audit from July 2014 to September 2015 in accordance with generally accepted government auditing standards. Those standards require that we plan and perform the audit to obtain sufficient, appropriate evidence to provide a reasonable basis for our findings and conclusions based on our audit objectives. We believe that the evidence obtained provides a reasonable basis for our findings and conclusions based on our audit objectives.

BACKGROUND

DOT is working with the automobile industry, state and local transportation agencies, researchers, private sector stakeholders, and others to lead and fund research on connected vehicle technologies to enable safe wireless communications among vehicles, infrastructure, and travelers' personal communications devices.[9] Connected vehicle technologies include vehicle-to-vehicle (V2V) and vehicle-to-infrastructure (V2I) technologies:

- V2V technologies transmit data between vehicles to enable applications that can warn drivers about potential collisions.
  Specifically, V2V-equipped cars would emit data on their speed, position, heading, acceleration, size, brake status, and other data (referred to as the "basic safety message") 10 times per second to the on-board equipment of surrounding vehicles, which would interpret the data and provide warnings to the driver as needed. For example, drivers may receive a forward collision warning when their vehicle is close to colliding with the vehicle in front of them. V2V technologies have a greater range of detection than existing sensor-based crash avoidance technologies available in some new vehicles.[10] NHTSA is pursuing actions to require that vehicle manufacturers install the underlying V2V technologies that would enable V2V applications in

new passenger cars and light truck vehicles, and requested comment on this issue in an August 2014 Advanced Notice of Proposed Rulemaking.[11] We reported on V2V technologies in November 2013.[12] Thus, we are not focusing on these technologies in this report.

- Vehicle-to-infrastructure (V2I) technologies transmit data between vehicles and the road infrastructure to enable a variety of safety, mobility, and environmental applications. V2I applications are designed to avoid or mitigate vehicle crashes, particularly those crash scenarios not addressed by V2V alone, as well as provide mobility and environmental benefits. Unlike V2V, DOT is not considering mandating the deployment of V2I technologies.

V2I applications rely on data sent between vehicles and infrastructure to provide alerts and advice to drivers. For example, the Spot Weather Impact Warning application is designed to detect unsafe weather conditions, such as ice or fog, and notify the driver if reduced speed or an alternative route is recommended (see left side of figure 1). DOT is also investigating the development of V2I mobility and environmental applications. For example, the Eco-Approach and Departure at Signalized Intersections application alerts drivers of the most eco-friendly speed for approaching and departing signalized intersections to minimize stop-and-go traffic and idling (see right side of fig. 1), and eco-lanes, combined with eco-speed harmonization, would provide speed limit advice to minimize congestion and maintain consistent speeds among vehicles in dedicated lanes.

DOT is also pursuing the development of V2I mobility applications that are designed to provide traffic signal priority to certain types of vehicles, such as emergency responders or transit vehicles. In addition, other types of V2I mobility applications could capture data from vehicles and infrastructure (for example, data on current traffic volumes and speed) and relay real-time traffic data to transportation system managers and drivers. For example, after receiving data indicating vehicles on a particular roadway were not moving, transportation system managers could adjust traffic signals in response to the conditions, or alert drivers of alternative routes via dynamic message signs located along the roadway. In addition to receiving alerts via message signs, these applications could also allow drivers to receive warnings through on-board systems or personal devices. Japan has pursued this approach through its ITS Spot V2I initiative, which uses roadside devices located along expressways to simultaneously collect data from vehicles to allow traffic managers to identify congestion, while also providing information to drivers regarding upcoming congestion and alternative routes.[13]

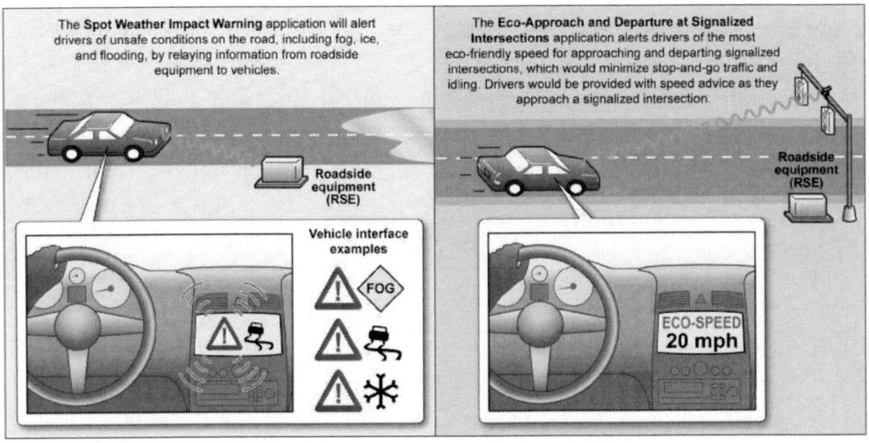

Source: GAO analysis of Department of Transportation documents. | GAO-15-775.

Figure 1. Examples of Vehicle-to-Infrastructure Applications.

To communicate in a connected vehicle environment, vehicles and infrastructure must be equipped with dedicated short-range communications (DSRC), [14] a wireless technology that enables vehicles and infrastructure to transmit and receive messages over a range of about 300 meters (nearly 1,000 feet).[15] As previously noted, V2Vequipped cars emit data on their speed, position, heading, acceleration, size, brake status, and other data (referred to as the "basic safety message") 10 times per second to the surrounding vehicles and infrastructure. V2I-equipped infrastructure can also transmit data to vehicles, which can be used by on-board applications to issue appropriate warnings to the driver when needed. According to DOT, DSRC is considered critical for safety applications due to its low latency,[16] high reliability, and consistent availability. In addition, DSRC also transmits in a broadcast mode, providing data to all potential users at the same time. Stakeholders and federal agencies have noted that DSRC's ability to reliably transfer messages between infrastructure and rapidly moving vehicles is an essential component to detecting and preventing potential collisions. DSRC technology uses radiofrequency spectrum to wirelessly send and receive data.[17] The Federal Communications Commission (FCC), which manages spectrum for nonfederal users, including commercial, private, and state and local government users, allocated 75 megahertz (MHz) of spectrum—the 5.850 to 5.925 gigahertz

(GHz) band (5.9 GHz band)[18]—for the primary purpose of improving transportation safety and adopted basic technical rules for DSRC operations.[19] However, in response to increased demands for spectrum,[20] FCC has requested comment on allowing other devices to "share" the 5.9 GHz band with DSRC technologies.[21]

V2I equipment may vary depending on the location and the type of application being used, although in general, V2I components in the connected vehicle environment include an array of roadside equipment (RSE) that transmits and receives messages with vehicles for the purpose of supporting V2I applications (see figure 2). For example, a V2Iequipped intersection would include:

- Roadside units (RSU)—a device that operates from a fixed position and transmits data to vehicles. This typically refers to a DSRC radio,[22] which is used for safety-critical applications that cannot tolerate interruption, although DOT has noted that other technologies may be used for non-safety-critical applications.
- A traffic signal controller that generates the Signal Phase and Timing (SPaT) message, which includes the signal phase (green, yellow, and red) and the minimum and maximum allowable time remaining for the phase for each approach lane to an intersection. The controller transfers that information to the RSU, which broadcasts the message to vehicles.
- A local or state back office, private operator, or traffic management center that collects and processes aggregated data from the roads and vehicles. As previously noted, these traffic management centers may use aggregated data that is collected from vehicles (speed, location, and trajectory) and stripped of identifying information to gain insights into congestion and road conditions as well.[23]
- Communications links (such as fiber optic cables or wireless technologies) between roadside equipment and the local or state back office, private operator, or traffic management center. This is typically referred to as the "backhaul network."
- Support functions, such as underlying technologies and processes to ensure that the data being transmitted are secure.

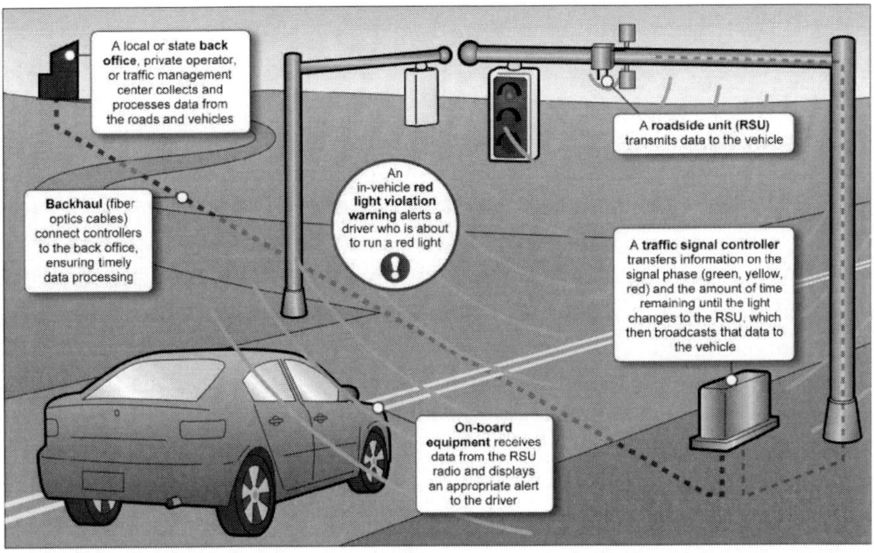

Source: GAO analysis of Department of Transportation documents. | GAO-15-775.

Figure 2. Example of Vehicle-to-Infrastructure Application Provided through Roadside Equipment.

V2I Deployment Efforts Are in the Early Stages and Extensive U.S. Deployment May Occur over the Next Few Decades

DOT and Various Stakeholders Are Developing and Testing V2I Technologies through Small Test Deployments

DOT, state and local transportation agencies, academic researchers, and private sector stakeholders are engaged in a number of efforts to develop and test V2I technologies and applications, as well as to develop the technology and systems that enable V2I applications. DOT's V2I work is funded through its connected vehicle research program. DOT's initial connected vehicle research focused on V2I technologies; however, it shifted its focus to V2V technologies because they are projected to produce the majority of connected vehicle safety benefits and they do not require the same level of infrastructure investment as V2I technologies. After conducting much of the research needed to inform its advanced notice of proposed rulemaking to require that vehicle manufacturers install V2V technologies in new passenger cars and light truck vehicles, DOT is now shifting its focus back to V2I technologies, and some of the technical work needed to develop V2V applications has also informed the development of V2I.[24] A number of DOT agencies are involved with the development and deployment of V2I technologies.[25] In addition, private companies have received contracts from DOT to develop the underlying concept of operations and technologies to support V2I applications, and auto manufacturers are collaborating with DOT in its efforts to develop and pilot certain V2I applications and the underlying technologies to support them. State and local transportation agencies, which will ultimately be deploying V2I technologies on their roads, have also pursued efforts to test V2I technologies in real-world settings. However, to date, only small research deployments (such as those described below) have occurred to test V2I technologies:

- **The Safety Pilot Model Deployment:** DOT partnered with the University of Michigan Transportation Research Institute to collect data to help estimate the effectiveness of connected vehicle technologies and their benefits in real-world situations. The pilot was conducted in Ann Arbor, Michigan, from August 2012 to February 2014, and included roughly 2,800 V2V-equipped cars, trucks, and buses, as well as roadside V2I equipment placed at 21 intersections, three curve-warning areas, and five freeway sites. While the primary focus was on V2V technologies, the pilot also evaluated V2I technology, such as Signal Phase and Timing (SPaT) technologies. DOT officials stated that it would be releasing six reports with findings from the Safety Pilot in mid to late 2015, although these reports will primarily focus on V2V applications. As of July 2015, DOT has released one report that included an evaluation of how transit bus drivers responded to V2V and V2I warnings, and of how well the test applications performed in providing accurate warnings.[26] The two V2I applications included were a curve speed warning and a warning that alerts the bus driver if pedestrians are in the intended path of the bus when it is turning at an intersection.
- **Connected Vehicle Pooled Fund Study:** A group of state transportation agencies, with support from the FHWA, established the Connected Vehicle Pooled Fund Study.[27] The study aims to aid transportation agencies in justifying and promoting the large scale deployment of a connected vehicle environment and applications through modeling, development, engineering, and planning activities.

To achieve this goal, the study funds projects that facilitate the field demonstration, deployment, and evaluation of connected vehicle infrastructure and applications. For example, the University of Arizona and the University of California at Berkeley are collaborating on a project to develop and test an intelligent traffic-signal system that could, among other things, provide traffic signal priority for emergency and transit vehicles, and allow pedestrians to request for more time to cross the street.

- **Crash Avoidance Metrics Partners, LLC (CAMP):** CAMP—a partnership of auto manufacturers that works to accelerate the development and implementation of crash avoidance countermeasures—established a V2I Consortium that focuses on addressing the technical issues related to V2I. In 2013, DOT awarded a cooperative agreement to CAMP, with a total potential federal share of $45 million, to develop and test V2I safety, mobility, and environmental applications, as well as the underlying technology needed to support the applications, such as security and GPS-positioning technologies. According to an FHWA official, CAMP's current efforts include developing, testing, and validating up to five V2I safety applications, as well as a prototype for Cooperative Adaptive Cruise Control, an application that uses V2V and V2I technology to automatically maintain the speed of and space between vehicles. In addition to CAMP, automakers have established the Vehicle Infrastructure Integration Consortium, which coordinates with DOT on connected vehicle policy issues, such as interoperability of V2I technologies.
- **Test Beds**: DOT, state and local agencies, and universities have established connected vehicle test beds. Test beds provide environments (with equipped vehicles and V2I roadside equipment) that allow stakeholders to create, test, and refine connected vehicle technologies and applications. This includes DOT's Southeast Michigan Test Bed, which has been in operation since 2007 to provide a real-world setting for developers to test V2I and V2V concepts, applications, technology, and security systems. In addition, state agencies and universities have established their own test beds. For example, the University Transportation Center in Virginia, in collaboration with the Virginia Department of Transportation, established the Northern Virginia Test Bed to develop and test V2I applications, some of which target specific problems—like

congestion—along the I-66 corridor. DOT offers guidance on how research efforts can become DOT-affiliated test beds, with the goal of enabling test beds to share design information and lessons learned, as well as to create a common technical platform. According to DOT, there are over 70 affiliated test bed members. The deployment of connected vehicle infrastructure to date has been conducted in test beds in locations such as Arizona, California, Florida, Michigan, New York, and Virginia.[28] Additionally, officials from some of these test beds told us they may apply to the Connected Vehicle Pilot Deployment Program later this year (see below).

- **The Connected Vehicle Pilot Deployment Program:** Over the next 5 years, DOT plans to provide up to $100 million in funding for a number of pilot projects that are to design and deploy connected vehicle environments (comprised of various V2I and V2V technologies and applications) to address specific local needs related to safety, mobility, and the environment. As envisioned, there are to be multiple pilot sites with each site having different needs, purposes, and applications.[29] The program solicitation notes that successful elements of the pilot deployments are expected to become permanent operational fixtures in the real-world setting (rather than limited to particular testing facilities), with the goal of creating a foundation for expanded and enhanced connected vehicle deployments. FHWA solicited applications for the pilot program from January through March 2015. According to DOT, the initial set of pilot deployments (Wave 1 award) is expected to begin in Fall 2015, with a second set (Wave 2 award) scheduled to begin in 2017. Pilot deployments are expected to conclude in September 2020.

DOT and Stakeholders Are Collaborating and Developing V2I Guidance for State and Local Agencies

DOT and other stakeholders have worked to provide guidance to help state and local agencies pursue V2I deployments, since it will be up to state and local transportation agencies to voluntarily deploy V2I technologies.[30] In September 2014, FHWA issued and requested comment on draft V2I deployment guidance intended to help transportation agencies make appropriate V2I investment and implementation decisions. For example, the guidance includes information on planning deployments, federal funding that can be used for V2I equipment and operations, technical requirements for equipment and systems, and applicable regulations, among other things. FHWA is updating the guidance and creating complementary guides, best practices, and toolkits, and officials told us they expect the revised guidance to be released by September 2015.[31] In addition, the American Association of State Highway and Transportation Officials (AASHTO),[32] in collaboration with a number of other groups, developed the *National Connected Vehicle Field Infrastructure Footprint Analysis*.[33] This report provides a variety of information and guidance for state and local agencies interested in V2I implementation, including a description of benefits; various state/local based scenarios for V2I deployments; underlying infrastructure and communications needs; timelines and activities for deployment; estimated costs and workforce requirements; and an identification of challenges that need to be addressed. AASHTO, with support the Institute of Transportation Engineers[34] and the Intelligent Transportation Society of America,[35] is also leading a V2I Deployment Coalition. The Coalition has several proposed objectives: support implementation of FHWA V2I deployment guidance; establish connected vehicle deployment strategies, and support standards development. According to information from the coalition and DOT, the V2I Deployment Coalition will be supported by technical teams drawn from DOT, trade associations, transportation system owners/operators, and auto manufacturers.

Extensive Deployment of V2I Technologies May Occur over the Next Few Decades

While early pilot-project deployment of V2I technologies is occurring, V2I technologies are not likely to be extensively deployed in the United States for the next few decades. According to DOT, V2I technologies will likely be slowly deployed in the United States over a 20-year period as existing infrastructure systems are replaced or upgraded. DOT has developed a connected vehicle path to deployment that includes steps such as releasing the final version of FHWA's V2I deployment guidance for state and local transportation agencies (September 2015),[36] and awarding and evaluating the Connected Vehicle Pilot Deployment Program projects in two phases, with the first phase of awards occurring in September 2015 and evaluation occurring in 2019, and the second phase of awards occurring in September 2017 and evaluation occurring in 2021. In addition, DOT officials noted that V2I will capitalize on V2V, and its deployment will lag behind the V2V rulemaking.

NHTSA will issue a final rule specifying whether and when manufacturers will be required to install V2V technologies in new passenger cars and light trucks.[37] In addition, FCC has not made a decision about whether spectrum used by DSRC can be shared with unlicensed devices, which could affect the time frames for V2I deployment. Even after V2I technologies and applications have been developed and evaluated through activities such as the pilot program, it will take time for state and local transportation agencies to deploy the infrastructure needed to provide V2I messages, and for drivers to purchase vehicles or equipment that can receive V2I messages. AASHTO estimated that 20 percent of signalized intersections will be V2Icapable by 2025, and 80 percent of signalized intersections would be V2Icapable by 2040. Similarly, AASHTO estimated that 90 percent of light vehicles would be V2V-equipped by 2040. However, DOT officials noted that environmental and mobility benefits can occur even without widespread market penetration and that other research has indicated certain intersections may be targeted for deployment. Similarly, in its *National Connected Vehicle Field Infrastructure Footprint Analysis*, AASHTO noted that early deployment of V2I technologies will likely occur at the highest-volume signalized intersections, which could potentially address 50 percent of intersection crashes.[38] See figure 3 for a list of planned events and milestones related to DOT's path to deployment of connected vehicle technologies.

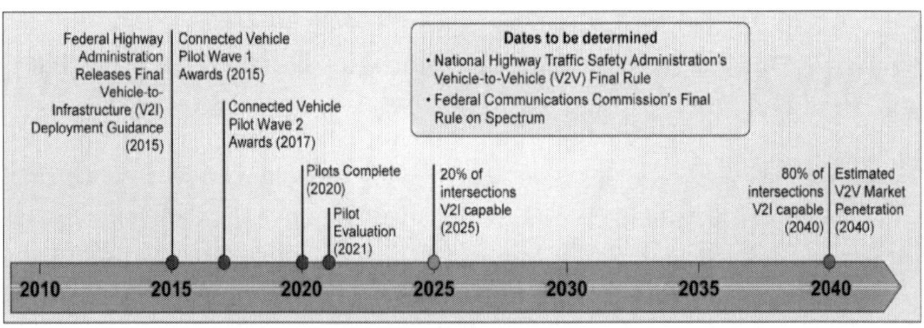

Figure 3. DOT's Planned Connected Vehicle Path to Deployment, 2010-2040.

A Variety of Challenges, Including Potential Spectrum Sharing, May Affect the Deployment of V2I

According to experts and industry stakeholders we interviewed, there are a variety of challenges that may affect the deployment of V2I technologies including: (1) ensuring that possible sharing with other wireless users of the radiofrequency spectrum used by V2I communications will not adversely affect V2I technologies' performance; (2) addressing states' lack of resources to deploy and maintain V2I technologies; (3) developing technical standards to ensure interoperability between devices and infrastructure; (4) developing and managing a data security system and addressing public perceptions related to privacy; (5) ensuring that drivers respond appropriately to V2I warnings; and (6) addressing the uncertainties related to potential liability issues posed by V2I. DOT is collaborating with the automotive industry and state transportation officials, among others, to identify potential solutions to these challenges.

Potential Spectrum Sharing

As previously noted, V2I technologies depend on radiofrequency spectrum, which is a limited resource in high demand due in part to the increase in mobile broadband use. To address this issue, the current and past administrations, Congress, FCC, and others have proposed a variety of policy, economic, and technological solutions to support the growing needs of businesses and consumers for fixed and mobile broadband communications by providing access to additional spectrum.[39] One proposed solution, introduced in response to requirements in the Middle Class Tax Relief and Job Creation Act of 2012,[40] would allow unlicensed devices[41] to share the 5.9 GHz band radiofrequency spectrum that had been previously set aside for the use of DSRC-based ITS applications such as V2I and V2V technologies.[42] FCC issued a Notice of Proposed Rulemaking in February 2013 that requested comments on this proposed solution.[43]

DOT officials and 17 out of 21 experts we interviewed considered the proposed spectrum sharing a significant challenge to deploying V2I technologies.[44] DSRC systems support safety applications that require the immediate transfer of data between entities (vehicle, infrastructure, or other platforms). According to DOT officials, delays in the transfer of such data due to harmful interference from unlicensed devices may jeopardize crash avoidance capabilities. Experts cited similar concerns, with one state official saying that if they deploy applications and they do not work due to harmful interference, potential users may not accept V2I. Seven experts we interviewed agreed that further testing was needed to determine if sharing would result in harmful interference to DSRC. In addition, DOT officials noted that changing to a shared 5.9 GHz band could impact current V2I research, which is based on the assumption that DSRC systems will have reliable access to the 5.9 GHz wireless spectrum.

According to Japanese government officials we interviewed, Japan also considered whether to share its dedicated spectrum with unlicensed devices and decided not to allow sharing of the spectrum used for V2I in the 700 MHz band.[45] According to officials we interviewed, Japan's Ministry of Internal Affairs and Communications conducted a study to test interference with V2I technologies and mobile phones to determine the impact on reliability and latency in delivering safety messages.[46] Based on these tests, the Japanese government decided not to allow sharing of the spectrum band used for V2I, because sharing could lead to delays or harmful interference with V2I messages. Japanese auto manufacturers we interviewed in Japan supported the decision of the Japanese government to keep the 700 MHz band dedicated to transportation safety uses. According to officials, if latency problems affect the receipt of safety messages, this could degrade the public's trust, consequently slowing down acceptance of the V2I system in Japan.

Since the Notice of Proposed Rulemaking was announced, various organizations have begun efforts to evaluate potential spectrum sharing in the 5.9 GHz band and some have expressed concerns. For example, harmful interference from unlicensed devices sharing the same band could affect the speed at which a V2I message is delivered to a driver. NTIA, which has conducted a study on the subject, identified risks associated with allowing unlicensed devices to operate in the 5.9 GHz band, and concluded that further work was needed to determine whether and how the risks identified can be mitigated. DOT also plans to evaluate the potential for unlicensed device interference with DSRC as discussed below.[47]

Given the pending FCC rulemaking decision, DOT, technology firms, and car manufacturers have taken an active role pursuing solutions to spectrum sharing. Specifically, DOT's fiscal year 2016 budget request included funds for technical analysis to determine whether DSRC can coexist with the operation of unlicensed wireless services in the same radiofrequency band without undermining safety applications.[48] According to DOT officials, since industry has not yet developed an unlicensed device capable of sharing the spectrum, the agency does not have a specific date for completion of this testing at this time. DOT officials noted, however, that they would work with NTIA in any spectrum-related matter to inform FCC of its testing results.[49] According to FCC officials we spoke with, FCC is currently collecting comments and data from government agencies, industry, and other interested parties and will use this information to inform their decision. For example, since 2013, representatives from Toyota, Denso, CSR Technology, and other firms worked together as part of the Institute of Electrical and Electronics Engineers (IEEE) DSRC Tiger Team[50] to evaluate potential options and technologies that would allow unlicensed devices to use the 5.9 GHz band without causing harmful interference to licensed devices. However, the representatives did not reach an agreement on a unified spectrum-sharing approach. Another ongoing effort from Cisco Systems, the Alliance of Automobile Manufacturers, and the Association of Global Automakers is preparing to test whether unlicensed devices using the "listen, detect and avoid"[51] protocol would be able to share spectrum without causing harmful interference to incumbent DSRC operations.[52] As of September 2015, FCC has not announced a date by which it will make a decision.

Lack of State and Local Resources to Develop and Maintain V2I Systems

Because the deployment of V2I technologies will not be mandatory, the decision to invest in these technologies will be up to the states and localities that choose to use them as part of their broader traffic-management efforts. However, many states and localities may lack resources for funding both V2I equipment and the personnel to install, operate, and maintain the technologies. In its report on the costs, benefits, and challenges of V2I deployment by local transportation agencies, the National Cooperative Highway Research Program (NCHRP) noted that many states they interviewed said that their current state budgets are the leanest they have been in years.[53] Furthermore, states are affected because traditional funding sources, such as the Highway Trust Fund, are eroding, and funding is further complicated by the federal government's current financial condition and fiscal outlook.[54] Consequently, there can be less money for state highway programs that support construction, reconstruction, and improvement of highways and bridges on eligible federal-aid highway routes, as well as for other authorized purposes.[55] According to one stakeholder we interviewed, there have been widespread funding cuts for state DOTs, and many state DOTs must first focus on maintaining the infrastructure and equipment they already have before investing in advanced technologies.[56] Ten experts we interviewed, including six experts from state and local transportation agencies, agreed that the lack of state and local resources will be a significant challenge to deploying V2I technologies.[57]

According to one report, without additional federal funding, deploying V2I systems would be difficult.[58]

Even if states decide to invest in V2I deployment, states and localities may face difficulties finding the resources necessary to operate and maintain V2I technologies. We have previously found that effectively using intelligent transportation systems, like V2I, depends on agencies' having the staff and funding resources needed to maintain and operate the technologies.[59] However, a recently released DOT report noted that staffing and information technology resources for maintaining V2I technologies were lacking in most agencies due to low and uncompetitive wage rates and funding constraints at the state and local government levels.[60] Similarly, 12 experts we interviewed stated that states and localities generally lack the resources to hire and train personnel with the technical skills needed to operate and maintain V2I systems.

According to FHWA's draft guidance on V2I deployment, funds are available for the purchase and installation of V2I technologies under various Federal-aid highway programs.[61] In addition, costs that support V2I systems, including maintenance of roadside equipment and related hardware, are eligible in the same way that other Intelligent Transportation System (ITS) equipment and programs are eligible. According to DOT, states have the authority and responsibility to determine the priority for funding V2I systems along with other competing transportation programs.

Japan's V2I systems, which were also voluntarily deployed, were funded in large part by the national government. According to Japan's National Police Agency, half of the costs for traffic signals were provided by the national government. In addition, according to the National Policy Agency, the Japanese government has invested an estimated $97 million (2014 dollars) in research and development for these systems. Two of the Japanese automakers we interviewed attributed the success of the Japanese V2I system in part to the significant government involvement and financial investment. Furthermore, according to a study on international connected vehicle technologies, Japan's nationally deployed and funded infrastructure devices allowed for industry partners to test and release connected vehicle technologies.[62]

Developing Technical Standards to Ensure Interoperability of V2I Systems

Nineteen of the 21 experts we spoke with reported that establishing technical standards is essential for all connected vehicle programs, including V2I, and will be challenging for a number of reasons.[63] According to DOT, such standards define how systems, products, and components perform, how they can connect, and how they can exchange data to interoperate.[64] DOT further noted that these standards are necessary for connected vehicle technologies to work on different types of vehicles and devices to ensure the integrity and security of their data transmission. As well, current standardization efforts have focused on standardizing the data elements and message sets that are transmitted between vehicles and the infrastructure.[65] Currently, according to DOT officials, DOT and various organizations have worked with the Society for Automotive Engineers (SAE) International to standardize the message sets and associated performance requirements for DSRC (SAE J2735 and J2945), which support a wide variety of V2V and V2I applications.[66] DOT, SAE International, and engineers from auto manufacturers, V2I suppliers, technology firms, and other firms meet to develop high-quality, safe, and cost-effective standards for connected vehicle devices and technologies, according to an expert from a leading industry organization specializing in setting connected vehicle technical standards. This expert also noted that developing consensus around what standards should be instituted could be difficult given the different interests (political, economic, or industry-related) of the many stakeholders involved in developing and deploying V2I technologies. For example, the expert said that developing effective security standards required for these technologies that are also cost-effective for auto manufacturers and government organizations to implement may be difficult.

Without common standards, V2I technologies may not be interoperable. DOT has noted that consistent, widely applicable standards and protocols are needed to ensure V2I interoperability across devices and applications. However, ensuring interoperability with a standard set of V2I applications in each state may be particularly challenging because unlike V2V, deployment of V2I technologies will remain voluntary. Consequently, states and localities may choose to deploy a variety of different V2I technologies—or no technologies at all based on what they deem appropriate for their transportation needs. DOT officials we interviewed recognized that a complete national deployment[67] of V2I technologies may never occur, resulting in a patchwork deployment of different applications in localities and states, although these applications will be required to be interoperable with one another. As a result, V2I deployment may be challenged by the following limitations:

- Benefits may not be optimized: Four experts we interviewed said that having a standard set of V2I applications in each state would be beneficial for drivers because a consistent deployment of applications could potentially increase benefits.
- Development of applications may be more limited: AASHTO's National Connected Vehicle Footprint Analysis argues that the more connected vehicle infrastructure is deployed nationwide using common standards, the more likely applications will be developed to take advantage of new safety, mobility, and environmental opportunities.
- Drivers may not find the system valuable: One expert from a state agency said without a standard set of V2I applications that allows drivers to use V2I applications seamlessly as they travel from state to state, travelers may lose confidence in the usefulness of the system and choose not to use it.

DOT and standardization organizations, such as the Society of Automotive Engineers (SAE) International, are working to develop standards to support DSRC and other V2I communications technologies. The data elements and message sets specified in the SAE standards are suitable not only for use with DSRC but also with other communications technologies such as cellular. According to DOT officials, the department is providing funding support, expert participation, and leadership in multiple standards development organizations to promote consensus on the key standards required to support nationally interoperable V2I and V2V technology deployments. Furthermore, the V2I Deployment Coalition—which includes AASHTO, the Institute of Electrical and Electronic Engineers, and the Institute of Transportation Engineers— intends to lead the effort to develop and support publishing of V2I standards, guidelines, and test specifications to support interoperability.[68] To facilitate standardization among potential state users of V2I technologies, FHWA is currently developing deployment guidance as discussed previously. According to DOT, that guidance will include specifications to ensure interoperability and to assist state and local agencies in making appropriate investment and implementation decisions for those agencies that will deploy, operate, and maintain V2I systems.

In addition to developing V2I standards across the United States, five experts we interviewed mentioned the importance of international harmonization for V2I technologies. Auto manufacturer experts recognized the importance of developing standards at both a domestic and international level as cars are manufactured globally. However, this is a challenge because international standardization organizations, including those in Europe and Japan, have different verification and validation processes than the United States, according to an auto manufacturer expert.[69] Furthermore, another expert noted that harmonization of standards is dependent on the country's or regional government's regulations, and since there are different views on the role of these regulations in Europe, Japan, and the United States, achieving global standards will be complex. According to DOT, the joint standardization of connected vehicle systems (V2V and V2I) is a core objective of European Union-U.S. cooperation on ITS, and U.S.-Japan staff exchanges have been invaluable in building relationships and facilitating technical exchange, thus creating a strong foundation for ongoing collaboration and research. According to DOT officials, even when identical standards are not viable across multiple countries or regions due to technical or legal differences, maximizing similarities can increase the likelihood that common hardware and software can be used in multiple markets, reducing costs and accelerating deployment. According to officials from one Japanese auto manufacturer we interviewed, developing a standard message set for V2I communications in Japan was a long and challenging process that took over 5 years of discussion among auto manufacturers.

Data Security System and Privacy Concerns

According to DOT, for connected vehicle technologies to function safely, security and communications infrastructure need to enable and ensure the trustworthiness of messages between vehicles and infrastructure. The source of each message needs to be trusted and message content needs to be protected from outside interference or attacks on the system's integrity. A DOT study[70] we reviewed and the majority of the experts we interviewed noted that data security challenges exist and cited challenges that range from securing messages delivered to and from vehicle devices and infrastructure to managing security credentials and associated policies for accessing data and the system.[71] Fourteen of 21 experts we interviewed cited securing data as a significant challenge to the deployment of V2I technologies.[72] For example, experts from 5 states and one local agency that operated V2I test beds told us they were uncertain how vehicle and infrastructure data would be stored and secured for a larger deployment of V2I technologies because they have only tested V2I applications in limited, small-scale deployments.[73] Most of these experts were also unsure whether current data security efforts could be scalable to a larger deployment. According to DOT officials, they are currently researching this area.

DOT and industry have taken steps to develop a security framework for all connected vehicle technologies, including V2I. DOT, along with automakers from CAMP, are testing and developing the Security Credential Management System (SCMS) to ensure the basic safety messages are secure and coming from an authorized device. More than half of the experts we interviewed expressed a variety of concerns about (1) the SCMS system, including whether SCMS can ensure a trusted and secure data exchange and (2) who will ultimately manage the system.[74] To solicit input on these issues DOT launched a Request for Information in October 2014 to obtain feedback in developing the organizational and operating structure for SCMS.[75] In our previous work on V2V,[76] we found that as a part of its research on the security system, DOT had identified three potential models—federal, public-private, and private. We previously found that if a federal model were pursued, according to DOT, the federal government would likely pursue a service contract that would include specific provisions to ensure adequate market access, privacy and security controls, and reporting and continuity of services.[77] We also reported that under a public-private partnership, the security system would be jointly owned and managed by the federal government and private entities. At the time of our prior report, DOT officials stated that its legal authority and resources have led NHTSA to focus primarily on working with stakeholders to develop a viable private model, involving a privately owned and operated security-management provider.

According to DOT officials, the agency is expanding the scope of its planned policy research to enable the Department to play a more active leadership role in working with V2V and V2I stakeholders to develop and prototype a private, multi-stakeholder organizational model for a V2V SCMS. Officials said that such a model would ensure organizational transparency, fair representation of stakeholders, and permit the federal government to play an ongoing advisory role. A central component of the Department's planned policy research is the development of policies and procedures that could govern an operational SCMS, including minimum standards to ensure security and appropriately protect consumer privacy. Currently, NHTSA is reviewing comments on the management and organization for SCMS to inform its V2V Notice of Proposed Rulemaking, expected to be submitted for Office of Management and Budget review by the end of 2015. In addition, according to DOT's Connected Vehicle Pilot Deployment Program request for proposals, participating state and local agencies will utilize SCMS as a tool to support deployment security, which will allow states, local agencies, and private sector firms an opportunity to test capabilities in a real-world setting. Ultimately, when asked about the sufficiency of SCMS, almost half of the experts we interviewed (10 of 21) indicated they were confident that a secure system for V2I could be developed.[78]

According to FHWA, a secure system is essential to appropriately protect the privacy of V2I users. Nine of the experts identified privacy as a significant challenge for the deployment of V2I technologies. For example, the public may perceive that their personal information could be exposed or their vehicle could be tracked using connected vehicle technologies.[79] In a connected vehicle environment, various organizations—federal, state, and local agencies; academic organizations; and private sector firms—potentially may have access to data generated by V2I technologies in order to, for example, manage traffic and conduct research. DOT has taken some steps to mitigate security[80] and privacy concerns related to V2V and V2I technologies.

According to DOT officials, the safety message will be broadcast in a very limited range (approximately 300 meters) and will not contain any information that identifies a specific driver, owner or vehicle (through vehicle identification numbers or license plate or registration information). The messages transmitted by DSRC devices (such as roadside units) in support of V2V and V2I technologies also will be signed by security credentials that change on a periodic basis (currently expected to be every 5 minutes) to minimize the risk that a third party could use the messages as a basis for tracking the location or path of a specific individual or vehicle.

Additionally, with respect to V2I technologies, DOT officials, car manufacturers and V2I suppliers plan to incorporate privacy by design into V2I technologies.[81] Under this approach, according to DOT, V2I data will be aggregated, and anonymized. Also NHTSA is currently in the process of conducting a V2V privacy risk assessment and intends to publish a Privacy Impact Assessment in connection with its V2V Notice of Proposed Rulemaking, which is expected to include an analysis of data collected, transmitted, stored, and disclosed by the V2V system components and other entities in relation to privacy concerns. The Department expects the V2V privacy risk research and the Privacy Impact Assessment to influence the development of policies, including security and privacy policies with regard to V2I. Furthermore, according to DOT, its V2I Deployment Coalition also plans to identify privacy and data issues at the state and county level.

According to Japanese officials we interviewed from the Ministry of Land, Infrastructure, Transport, and Tourism (MLIT), Japan took a number of steps to address the security and privacy of its V2I system. First, Japan's Intelligent Transportation Systems Technology Enhancement Association is responsible for managing the security of their V2I systems, and developed a system that used encryption to maintain security and ensure privacy. More specifically, each vehicle participating in V2I is assigned a changing, random identification number each time the vehicle started, thus making it difficult to track the vehicle over time. MLIT officials also noted that data generated from each vehicle is not stored permanently, but rather saved for distinct time frames depending on its use. Further, MLIT officials stated that security is ensured because V2I information is protected, anonymous, non-identifiable, and not shared with outside organizations; rather, it is used solely for public safety purposes. According to the National Police Agency officials, no significant security issue has occurred with V2I technologies as of July 2015.

Human Factors

Because V2I data will initially provide alerts and warning messages to drivers, the ultimate effectiveness of these technologies, especially as it relates to safety, depends on how well drivers respond to the warning messages.[82] In a November 2013 report on V2V technologies, we found that addressing human factors that affect how drivers will respond included (1) minimizing the risk that drivers could become too familiar with or overly reliant upon warnings over time and fail to exercise due diligence in responding to them, (2) assessing the risk that warnings could distract drivers and present new safety issues, and (3) determining what types of warnings will maximize driver response.[83] Seven of the 21 experts we interviewed identified human factors issues as significant to V2I deployment.[84]

To address these concerns, DOT is participating in a number of research efforts to determine the effects of new technologies on driver distraction. To further examine the effects on drivers using V2I applications, NHTSA has a research program in place to develop human factors principles that may be used by automobile manufacturers and suppliers as they design and deploy V2I technology and other driver-vehicle interfaces that provide warnings to drivers.[85] In addition, DOT's ITS-JPO is funding NHTSA and FHWA research to investigate human factors implications for V2I technologies. Furthermore, according to DOT, the Connected Vehicle Pilot Program will allow additional opportunities to review drivers' reactions to V2I messages using cameras and driver vehicle data on speed, braking, and other metrics.

Liabilities

Eleven of the 21 experts we interviewed identified uncertainty related to potential liability in the event of a collision involving vehicles equipped with V2I technologies as a challenge.[86] In our November 2013 report on V2V, an auto manufacturer expert said that it could be harder to determine whether fault for a collision between vehicles equipped with connected vehicle technologies lies with one of the drivers, an automobile manufacturer, the manufacturer of a device, or another party.

According to DOT officials, it is unlikely that either V2I or V2V technologies will create significant liability exposure for the automotive industry, as DOT expects auto manufacturers will contractually limit their potential liability for integrated V2I and V2V applications and third-party services. However, according to DOT, V2I applications using data received from public infrastructure may create potential new liability risks to various infrastructure owners and operators—state and local governments, railroads, bridge owners, and roadway owners—because such cases often are brought against public or quasi-public entities and not against vehicle manufacturers. According to DOT, this liability will likely be the same as existing liability for traffic signals and variable message signs.

Extent of Benefits and Costs Are Likely to Remain Unclear until Further Deployment of V2I Technology

Experts Identified Potential Safety, Mobility, Operational, and Environmental Benefits to V2I Technologies, but Extent of Benefits Is Not Yet Clear

DOT officials, stakeholders representing state officials and private sector entities, and experts we interviewed stated that the deployment of V2I technologies and applications is expected to result in a variety of benefits to users. Experts identified safety, mobility, operational, and environmental benefits as the potential benefits of V2I.

- Safety: Eleven of 21 experts identified safety as one of the primary benefits of V2I technologies.[87] This included 6 of the 8 state and local agencies we interviewed. According to Japanese officials we interviewed, Japan has realized safety benefits from its deployment of V2I infrastructure. For example, in an effort to prevent rear-end collisions, Japan installed V2I infrastructure that detected and warned motorists of upcoming congestion on an accident-prone curve on an expressway in Tokyo. According to Japanese officials, this combined with other measures such as road marking, led to a 60-percent reduction in rear-end collisions on this curve.
- Mobility: In interviews, 8 of 21 experts identified mobility as one of the primary benefits of V2I, including 6 of the 8 state and local agencies we interviewed.[88] Officials in three states we interviewed noted that they are focusing on V2I applications that have the potential to increase mobility. These applications could allow for transportation system managers to identify and address congestion in real-time, as well as provide traffic signal priority to certain types of vehicles, such as emergency responders or transit. For example, Japanese officials estimated that as the use of electronic tolling rose to nearly 90 percent of vehicles on expressways, tollgate congestion was nearly eliminated on certain expressways.[89]
- Operations: In interviews, 7 of 21 experts, including 4 of 8 state and local agencies, identified the potential for V2I applications to provide operational benefits or cost savings.[90] For example, one state agency

noted that using data collected from vehicles could allow the transportation managers to more easily monitor pavement conditions and identify potholes (typically a costly and resource-intensive activity). DOT and the National Cooperative Highway Research Program[91] have also noted that the visibility and enhanced data on current traffic and road conditions provided by V2I applications would provide operational benefits to state and local transportation managers. This result, in turn, could provide safety or other benefits to drivers. For example, officials in Japan told us that by using data collected from vehicles through the ITS infrastructure, they were able to identify 160 locations in which drivers were braking suddenly. After investigating the cause, officials took steps to address safety issues at these sites (such as trimming trees that created visual obstructions) and incidents of sudden braking decreased by 70 percent and accidents involving injuries or fatalities decreased 20 percent. In addition, the Japanese government partnered with private industry to collect and analyze vehicle probe data to help the public determine which roads were passable following an earthquake.

- Environment: Of the experts we interviewed, 4 of 21 identified environmental benefits as a primary benefit of V2I technologies, with some noting interconnections among safety, mobility, and environmental benefits.[92] For example, officials from two state agencies we interviewed stated that improving safety and mobility will lead to environmental benefits because there will be less stop-and-go traffic. Indeed, Japanese officials estimated that decreased tollgate congestion reduced CO2 emissions by approximately 210,000 tons each year.

Although V2I applications are being developed for the purpose of providing safety, mobility, operational, and environmental benefits, the extent to which V2I benefits will be realized is currently unclear because of the limited data available and the limited deployment of V2I technologies. To date, only small research deployments have occurred to test connected vehicle technologies. However, DOT has commissioned or conducted some studies to estimate potential V2I benefits, particularly with respect to safety and the environment.

- NHTSA used existing crash data and estimated that in combination, V2V and V2I could address up to 81 percent of crashes involving unimpaired drivers.[93] Similarly, in 2012, a study commissioned by FHWA used existing crash data and estimated the number, type, and costs of crashes that could be prevented by 12 different V2I applications. This study estimated that the 12 V2I applications would prevent 2.3-million crashes annually (representing 59 percent of single vehicle crashes and 29 percent of multi-vehicle crashes and comprising $202 billion in annual costs).[94]
- With respect to the environment, DOT contracted with Booz Allen Hamilton to develop an initial benefit-cost analysis for its environmental applications, with the goal of informing DOT's future work and prioritization of certain applications.[95] As part of the next phase of this work, Booz Allen Hamilton used models to estimate potential benefits of individual applications, as well as their benefits when used in combination with other applications.[96]
- NCHRP estimated operational and financial benefits that V2I applications may provide to state and local governments, such as reduced costs for crash response and cleanup costs; reduced need for traveler information infrastructure; reduction of infrastructure required to monitor traffic; and lower cost of pavement condition detection.[97] However, one of the study's major conclusions was that the data required to quantify benefits are generally not available.
- DOT is taking some steps to evaluate the benefits of V2I applications. For example, as part of its upcoming Connected Vehicle Pilot Deployment Program, pilot projects are expected to develop a performance-monitoring system, establish performance measures, and collect relevant data. Projects will also receive an independent evaluation of their projects' costs and benefits; user acceptance and satisfaction; and lessons learned.

In addition, organizations researching the benefits of V2I have noted that the benefits of V2I deployments may depend on a variety of factors, including the size and location of the deployment, the number of roadside units deployed, the number of vehicles equipped, and the types of applications that are deployed. A study sponsored by the University of Michigan Transportation Research Institute noted that some V2I safety applications require a majority of vehicles to be equipped before reaching optimum effectiveness, in contrast to mobility, road weather, and operations applications, which only require a small percentage of equipped vehicles before realizing benefits.[98] Japanese government officials, as well as representatives from a private company we interviewed in Japan, noted that in some cases, they have found it difficult to quantify benefits. However, DOT and the Ministry of Land, Infrastructure, Transport and Tourism of Japan established an Intelligent Transportation Systems (ITS) Task Force to exchange information and identify the areas for collaborative research to foster the development and deployment of ITS in both the United States and Japan. According to DOT, evaluation tools and methods are high-priority areas for the task force, and DOT has stated that a report detailing the task force's collaborative research on evaluation tools and methods will be published in 2015. In addition, 8 of the 21 experts we interviewed noted that it can be difficult to identify benefits that are solely attributable to V2I, due to the interconnected nature of V2V and V2I technologies.[99] However, some experts we spoke with provided some examples of how connected vehicle benefits could be measured, including: crash avoidance, reduction in fatalities, reduced congestion, and reduced travel times.

The Costs for V2I Are Unclear due to Limited Deployment

The costs for the deployment of a national V2I system are unclear because current cost data for V2I technology are limited due to the small number of test deployments thus far. According to DOT officials, experts, and other industry stakeholders we spoke to, there are two primary resources for estimating V2I deployment costs: AASHTO's *National Connected Vehicle Footprint Analysis* (2014) and National Cooperative Highway Research Program's (NCHRP)[100] *03-101 Costs and Benefits of Public-Sector Deployment of Vehicle-to-Infrastructure Technologies* (2013).[101] However, the cost estimates in both reports are based on limited available data from small, research test beds. As a result, neither report contains an estimate for the total cost if V2I were to be deployed at a national level.[102] Despite these limitations, the cost estimates in these two studies are cited by several experts and industry stakeholders, including DOT.[103] According to DOT, these cost figures may be useful to agencies considering early deployments.

According to AASHTO and NCHRP, costs of V2I deployment will likely be comprised of two types of costs. First, V2I will require *non-recurring* costs—the upfront, initial costs required to deploy the infrastructure.

According to AASHTO, there are two primary, non-recurring cost categories associated with V2I deployments:

- *Infrastructure deployment* costs include the costs for planning, acquiring, and installing the V2I roadside equipment. State and local agencies will need to evaluate the costs for planning and design that may include mapping intersections and deciding where to deploy the DSRC radios based on traffic and safety analyses, according to AASHTO. Deployment costs will include the cost of acquiring the equipment, including the roadside unit. AASHTO estimates that the total equipment costs would be $7,450 per site,[104] with $3,000 attributed to each roadside unit, on average.[105] However, 4 of the experts we interviewed stated that the cost estimates for the hardware are likely to decrease over time, as the technology matures and the market becomes more competitive. The total average cost for installation of the equipment per site includes the costs of labor and inspection. In addition, deployment costs may include the cost of upgrading traffic signal controllers.[106] AASHTO estimates that approximately two thirds of all controllers in the United States will need to be upgraded to support connected vehicle activities.
- *Backhaul costs* refer to the costs for establishing connectivity for communication between roadside units and back offices or traffic management centers (TMCs).[107] As discussed, backhaul includes the fiber optic cables connecting traffic signals to the back office, as well as any sensors or relays that link to or serve these components. According to NCHRP, backhaul will be one of the biggest components of costs. In fact, three state agencies and one supplier we spoke with referred to backhaul as a factor that will affect costs for V2I deployment. Backhaul costs are also uncertain because states vary in the extent to which they have existing backhaul. According to AASHTO, some sites may only require an upgrade to their current backhaul system to support expected bandwidth requirements for connected vehicle communications. However, 40 percent of all traffic signals have either no backhaul or will require new systems, according to AASHTO. The difference in cost between tying into an existing fiber-optic backhaul and installing a new fiber-optic backhaul for the sites is significant, according to DOT. The average national cost to upgrade backhaul to a DSRC roadside site is estimated to vary from $3,000, if a site has sufficient backhaul and will only need an

upgrade, to $40,000, if the V2I site requires a completely new backhaul system, according to AASHTO estimates.

The total potential average, non-recurring costs of deploying connected vehicle infrastructure per site, according to DOT and AASHTO, are $51,650 (see table 1).[108]

Table 1. Total Potential Average, Non-Recurring Costs of Connected Vehicle Infrastructure per Site

Cost element	Average cost
Planning and design average costs	$6,650
Equipment average costs	$7,450
Installation average costs	$3,550
Backhaul average costs	$30,800
Signal controller upgrades average costs	$3,200
Total potential average costs per site	$51,650

Source: American Association of State Highway and Transportation Officials and DOT, | GAO-15-775 Note: "Per site" refers to each specific location where a roadside unit is deployed.

Second, V2I will also require recurring costs—the costs required to operate and maintain the infrastructure. According to AASHTO, there are several types of recurring costs associated with V2I deployments, including equipment maintenance and replacement, security, and personnel costs. The amount of maintenance needed to keep roadside units running is unclear, according to 3 of the experts we interviewed, because the test bed deployments have generally not operated long enough to warrant maintenance of the equipment. However, NCHRP estimates that routine maintenance costs for roadside units would likely vary from 2 to 5 percent of the original hardware and labor costs. This includes such maintenance as realigning antennas and rebooting hardware. AASHTO also estimates that the device would need replacing every 5 to 10 years. In addition, states and localities may also need to hire new personnel or train existing staff to operate these systems. According to AASHTO, personnel costs will also depend on the size of the deployment as smaller deployments may not need dedicated personnel to complete maintenance, while large deployments may require staff dedicated to system monitoring on site or on call. Furthermore, security costs will be a recurring cost and include the costs of keeping the security credentials of the

SCMS up to date and the costs to manage the security system, according to AASHTO. Given that SCMS is still being developed, cost estimates are unknown.[109] One car manufacturer we interviewed explained that because the management of the security system is unknown, it is extremely challenging to estimate future costs. In addition, one county agency official said security costs could greatly affect the total costs for V2I deployment because the requirements and funding responsibility are not clearly defined. As part of its ANPRM, NHTSA conducted an assessment of preliminary V2V costs, including costs for the SCMS.[110] NHTSA estimated that the SCMS costs per vehicle range from $1 to $6, with an average of $3.14. SCMS costs will increase over time due to the need to support an increasing number of vehicles with the V2V technologies, according to NHTSA.

While AASHTO and NCHRP have estimated the above potential average costs for various components associated with a V2I deployment, 10 of 21 experts stated that it is difficult to determine the actual costs for a V2I deployment in a particular state or locality due to a number of factors.[111]

- First, the scope of the deployment will affect the total costs of a region's V2I deployment, according to NCHRP, because it will determine the amount of equipment needed for the system to function, including the number of roadside units. Previous test bed deployments have varied in size ranging from 1 to 2,680 DSRC roadside units. Further, the number of devices needed will be dependent on how many devices are required to enable the applications. For example, while a curve-speed-warning application may require installing equipment at a specific location, applications that aim to mitigate congestion by advising drivers of the best speed to approach an intersection may need to be installed at several intersections throughout an urban corridor. One state agency said that one factor that could affect costs is how often roadside equipment needs to be replaced in order to enable certain V2I applications. In addition, as previously mentioned, the size of the deployment will contribute to personnel costs.
- Second, the state or locality's deployment environment will affect its deployment costs. One state agency pointed out that everyone's costs will be different because they will be deploying in environments with differing levels of existing infrastructure. For example, as previously noted, the region's existing backhaul

infrastructure will determine the extent of the cost for installing or upgrading the region's system, including whether a city or state has fiber optics already installed or signal controllers need upgrading.

- Lastly, the maturity of the technology will also affect cost estimates for equipment such as a DSRC radio. Estimating equipment costs is difficult at this time because the technology is still developing, according to NCHRP. Ten of the 21 experts we interviewed, including all of the state agencies, also mentioned that estimating costs is challenging because the technology is still immature.[112] Furthermore, the reports and 4 experts we interviewed agree that the cost estimates for the hardware are likely to decrease over time, as the technology matures and the market becomes more competitive.[113]

As part of the upcoming Connected Vehicle Pilot Deployment Program, DOT developed the Cost Overview for Planning Ideas and Logical Organization Tool (CO-PILOT). This tool generates high-level cost estimates for 56 V2I applications based on AASHTO's estimations. In addition, according to DOT, the agency will work with AASHTO to develop a life-cycle cost tool that agencies can use to support V2I deployment beyond the Connected Vehicle Pilot Deployment Program. DOT officials also indicated that they plan to update the tool over time as more data are collected from the Connected Vehicle Pilot Deployment Program, and they expect the tool to be available for use by 2016. Also, as previously mentioned, FHWA is developing deployment guidance that will outline potential sources of funding for states and localities, among other things.

AGENCY COMMENTS

We provided a draft of this product to the Secretary of Transportation, Secretary of Commerce, and the Chairman of the FCC, for review and comment. DOT and Commerce's NTIA both provided comments via email that were technical in nature. We incorporated these comments as appropriate. FCC did not provide comments.

David Wise
Director, Physical Infrastructure

Appendix I: Scope and Methodology

To address all of our objectives, we reviewed documentation relevant to vehicle-to-infrastructure (V2I) technology research efforts of the Department of Transportation (DOT), state and local government, and automobile industry, such as DOT's 2015 Federal Highway Administration V2I Draft Deployment Guidance and Products and AASHTO's *National Connected Vehicle Field Infrastructure Footprint Analysis*, as well as documentation on completed and ongoing research. We interviewed officials from DOT's Office of the Assistant Secretary for Research and Technology, Intelligent Transportation Systems-Joint Program Office (ITS-JPO), Federal Highway Administration (FHWA), National Highway Traffic Safety Administration (NHTSA), and the Volpe National Transportation Systems Center, about these efforts.

For all objectives, we developed a structured set of questions for our interviews with 21 experts who represented domestic automobile manufacturers, V2I device suppliers, state and local government, privacy experts, standardization organizations, and academic researchers with relevant expertise. The identified experts have varying degrees of expertise in the following areas related to V2I technology: the production of passenger vehicles; technology development; technology deployment; data privacy; security; state agency deployment; and legal and policy issues. Our starting point for our expert selection was a list of experts originally created in January 2013 by the National Academy of Sciences for GAO's vehicle-to-vehicle (V2V) report.[1] We used this list for our initial selection because V2V and V2I technologies are both connected vehicle technologies with many similarities, and many V2V stakeholders are also working on V2I.[2] In addition to nine experts we selected from the National Academy of Sciences list, we selected an additional 12 experts based on the following factors:

1) their personal involvement in the deployment of V2I technologies;
2) recommendations from federal agencies (DOT, and the Federal Communications Commission (FCC) and associations (such as the American Association of State and Highway Transportation Officials (AASHTO); and
3) experts' involvement in professional affiliations such as a V2I consortium or groups dedicated to these technologies or to a specific challenge affecting V2I (e.g., privacy).

Table 2 lists the experts we selected.

In conducting our structured interviews, we used a standardized interview to ensure that we asked all of the experts the same questions. During these interviews we asked, among other things, for expert views on the state of development and deployment of V2I technologies (including DOT's role in this process), the potential benefits of V2I technologies, and their potential costs. We also asked for each expert's views on a number of defined potential challenges facing the deployment of V2I technologies, and asked the experts to rate the significance of each challenge using a three-point scale (significant challenge, moderate challenge, or slight challenge). We determined this list of potential challenges after initial interviews with DOT, industry associations, and other interest groups knowledgeable about V2I technologies. Prior to conducting the interviews, we tested the structured interview with one association to ensure our questions were worded appropriately. After conducting these structured interviews, we summarized expert responses relevant to each objective. The viewpoints gathered through our expert interviews represent the viewpoints of individuals interviewed and cannot be generalized to a broader population.

For the purpose of this review, state and local agency officials were considered experts because of their experience in deploying and testing V2I technologies, and experience working with the required technologies (DSRC equipment and software), decision process (funding and scheduling); personnel requirements and skill sets needed for deployment; operations and maintenance. We specifically included six officials who deployed V2I test beds in their respective states in our pool of expert interviews.

We also included two officials who studied V2I for several years, had taken part in the AASHTO's Connected Vehicle group, and had applied to DOT's prior Connected Vehicles Pilot Program (V2I test bed). We also interviewed additional officials who have contributed to the U.S. efforts to develop and deploy connected vehicle technologies—officials who we refer to as "stakeholders." Specifically, we used these stakeholders to help us understand issues that informed our structured set of questions, but did not administer the structured question set during these stakeholder interviews. We primarily selected stakeholders based on recommendations from DOT and industry associations. However, we also included DOT as a stakeholder in the deployment of V2I technologies because it is leading federal V2I efforts.

Table 2. Subject Matter Experts Interviewed

Experts	Organization
Roger Berg	Denso
Debra Bezzina	University of Michigan Transportation Research Institute
Dorothy Glancy	Santa Clara University School of Law
Michael Shulman	Ford and Crash Avoidance Metrics Partners, LLC
Tom Schaffnit	Vehicle Infrastructure Integration Consortium
Ravi Puvvala	SavariNetworks
DavidMiller	Siemens
Barry Einsig	Cisco
Jack Pokrzywa	SAE International
Siva R.K.Narla	Institute of Transportation Engineers
Scott Geisler	General Motors
Steven Siko	Chrysler
Jules Polonetsky	Future for Privacy Forum
Greg Larson	California Department of Transportation
Richard McDonough	New York State Department of Transportation
Matt Smith	Michigan Department of Transportation
Catherine McGhee	Virginia Department of Transportation
Elizabeth Birriel	Florida Department of Transportation
Charlie Farnham	Texas Department of Transportation
Blaine Leonard	Utah Department of Transportation
Faisal Saleem	Maricopa County Department of Transportation, Arizona

Source: GAO. | GAO-15-775

We interviewed officials from 17 V2I stakeholder organizations including:

1) DOT, NHTSA
2) DOT, Office of the Assistant Secretary for Research and Technology, ITS-JPO
3) DOT, FHWA
4) DOT, Volpe National Transportation Systems Center
5) DOT, Chief Privacy Officer
6) National Telecommunications and Information Administration (NTIA)
7) FCC

8) AASHTO
9) Intelligent Transportation Society of America (ITS America)
10) Crash Avoidance Metrics Partners, LLC (CAMP)
11) Institute of Electrical and Electronics Engineers (IEEE)
12) National Cooperative Highway Research Program (NCHRP)
13) Leidos, previously known as Science Applications International Corporation (SAIC)
14) Virginia Tech Transportation Institute
15) Virginia Department of Transportation
16) Minnesota Department of Transportation
17) Road Commission for Oakland County, Michigan

To determine the status of development and deployment of V2I technology, we interviewed officials from DOT, including the Office of the Assistant Secretary for Research and Technology, ITS-JPO, FHWA, Volpe National Transportation Systems Center, and the NHTSA. We also interviewed officials at all seven V2I test beds located in Virginia, Michigan, Florida, Arizona, California, and New York.[3] We conducted site visits to three test beds—the Safety Pilot in Ann Arbor, Michigan, and the test beds in Southeast Michigan and Northern Virginia. We selected the three site visit locations based on which had the most advanced technology according to DOT and state officials. At these site visits, we conducted interviews with officials from state and local transportation agencies and academic researchers to collect information on developing and deploying V2I technology. We visited FHWA's Turner Fairbank Highway Research Center in Virginia to understand the agency's connected vehicle research efforts. We reviewed documentation of the efforts of DOT and automobile manufacturers related to vehicle-toinfrastructure (V2I) technologies, such as the 2015 FHWA's V2I *Draft Deployment Guidance and Products* and documentation on completed and ongoing research. We identified materials published in the past 4 years that were related to the terms "vehicle-to-infrastructure" and "V2I" through searches of bibliographic databases, including Transportation Research International Documentation and WorldCat. While a variety of V2I technologies exist for transit and commercial vehicles, for the purpose of this report we limited our scope to passenger vehicles since much of DOT's connected vehicle work is focused on passenger vehicles.

To determine the challenges affecting the deployment of V2I technology and DOT's existing or planned actions to address potential challenges, we reviewed FHWA's V2I draft guidance to assist in planning for future investments and deployment of V2I systems. In addition, we interviewed officials from FCC and NTIA about challenges related to the potential for spectrum sharing in the 5.9 GHz band. We interviewed DOT's Privacy Officer, two privacy experts, and several stakeholders to understand privacy concerns regarding the deployment of V2I technologies.

We collected information on anticipated benefits of these technologies through interviews with officials from DOT, automobile manufacturers, industry associations, and experts identified by National Academy of Sciences and other stakeholders, and through reviews of studies they provided. To specifically address the potential costs associated with V2I technologies, we analyzed two reports, AASHTO's *National Connected Vehicle Field Infrastructure Footprint Analysis* and NCHRP's 03-101, *Cost and Benefits of Public-Sector Deployment of Vehicle-to-Infrastructure Technologies* report, both of which addressed acquisition, installation, backhaul, operations, and maintenance costs.[4] According to DOT officials and other stakeholders we interviewed, those two reports were the primary sources of information for V2I potential deployment costs estimates and actual costs.[5] We used V2I costs estimates from the *AASHTO Footprint Analysis* to give examples of potential costs for deployment. To further assess the reliability of the cost estimates, in addition to our own review of the two reports, our internal economic stakeholder also independently reviewed both reports, and we subsequently interviewed representatives from AASHTO and NCHRP to verify the scope and methodology of the cost analyses performed in both reports. In addition, we discussed estimated costs and factors that affected costs for V2I investments with experts and stakeholders from federal, local, state government, academia, car manufacturers, industry associations, and V2I suppliers. We determined that the actual cost figures were reliable and suitable for the purpose of our report.

In addition to the above work, we selected Japan for a site visit because of its nationwide deployment and years of experience with deployment and maintenance of V2I technologies. Japan has led efforts in V2I technology development and deployment for over two decades. The country serves as an illustrative example from which to draw information on potential benefits, costs, and challenges of deploying V2I technologies in the United States.[6] During our site visit, we interviewed Japanese government officials and auto manufacturers on similar topics that we discussed with U.S. experts, including V2I deployment efforts, benefits, costs, and challenges.

Japanese Government

- Cabinet Secretariat (IT Strategy)
- Cabinet Office (Council for Science and Technology Policy)
- Ministry of Land, Infrastructure, Transport and Tourism (MLIT)
  - Road Bureau
  - Road Transport Bureau
- Ministry of Internal Affairs and Communications (MIC)
- National Police Agency (NPA)
- Ministry of Economy, Trade and Industry (METI)

Car Manufacturers

- Toyota
- Honda
- Nissan

V2I Supplier

- Denso

We conducted this performance audit from July 2014 through September 2015 in accordance with generally accepted government auditing standards. Those standards require that we plan and perform the audit to obtain sufficient, appropriate evidence to provide a reasonable basis for our findings and conclusions based on our audit objectives. We believe that the evidence obtained provides a reasonable basis for our findings and conclusions based on our audit objectives.

Appendix II: Expert Ratings of Potential Challenges Facing Deployment of Vehicle - to-Infrastructure Technologies

As part of our review, we conducted 21 structured interviews with individuals identified by the National Academy of Sciences and based on other factors discussed in our scope and methodology to be experts on vehicle-to-infrastructure (V2I) technologies (see table 2 in app. I for list of experts interviewed). During these interviews we asked, among other things, for each expert's views on a number of already defined potential challenges facing the deployment of V2I technologies. The ratings provided by the experts for each of the potential challenges discussed are shown in table 3 below. To inform our discussion of the challenges facing the deployment of V2I technologies, we considered these ratings as well as experts' responses to open-ended questions.

Table 3. Expert Ratings of Potential Challenges Facing Deployment of Vehicle-to-Infrastructure Technologies

Potential Challenges	Significant Challenge	Moderate Challenge	Slight Challenge	Did Not Answer
Spectrum sharing—allowing unlicensed devices to operate in the 5.9 GHz band along with V2I technologies	17	1	1	2
Non-mandatory, voluntary deployment of V2I technologies among states and localities	12	6	1	2
Limitations at the federal level for V2I deployment in terms of funding or providing technical resources to state/local transportation operators	9	2	3	7
Limitations at the state/local level for V2I deployment (for example, funding, personnel, or other areas)	10	6	0	5
Ensuring privacy under a system that involves the sharing of data among vehicles and government infrastructure.	9	5	6	1
Data Security measures are in place to secure the data that is collected at the Test Bed [Question for Test Bed locations only] [a] (Total = 7)	6	1	0	0
Ensuring data security in deploying V2I technologies.[Question for a non-test bed interviews—car manufacturers, associations, and others] [a] (Total = 14)	8	3	2	1
Standardization	7	8	5	1
Human factors in the deployment of V2I technologies	7	10	2	2
Liability issues, in terms of uncertainty related to legal responsibility for vehicle crashes using V2I technology	5	7	5	4

Source: GAO analysis of expert interviews. | GAO-15-775

[a]We asked separate data security questions depending on whether the expert had previously or currently worked on a V2I test bed.

End Notes

[1] Department of Transportation (DOT) National Highway Traffic Safety Administration (NHTSA), *Traffic Safety Facts 2013*, DOT HS 812 139 (Washington, DC).

[2] A variety of software applications are being developed that would use V2I technologies to provide different types of information to drivers.

[3] These are just a few examples of V2I applications: DOT has defined over 40 potential applications that would serve a wide range of functions.

[4] DOT, *2015 FHWA Vehicle to Infrastructure Deployment Guidance and Products* (Draft), version 9 (September 9, 2014).

[5] We primarily selected stakeholders based on recommendations from DOT and industry associations.

[6] There are two test beds in Michigan, the Safety Pilot in Ann Arbor, Michigan, and one in Southeast Michigan, in Oakland County.

[7] In conducting our structured interviews, we used a standardized interview guide to ensure that we asked all of the experts the same questions.

[8] In the 1990s, Japan introduced its Vehicle Information and Communication System (VICS), which provides real-time road traffic information to drivers via a VICS-equipped navigation device. In 2011, Japan implemented V2I with its deployment of ITS Spot. In 2010, DOT and Japan's Ministry of Land, Infrastructure, Transport, and Tourism (MLIT) signed a memorandum of cooperation to promote bilateral collaboration in the field of ITS.

[9] This effort is part of DOT's Intelligent Transportation Systems Program (ITS). ITS technologies consist of a range of communications, electronics, and computer technologies, such as systems that collect real-time traffic data and transmit information to the public via means such as dynamic message signs, ramp meters to improve the flow of traffic on freeways, and synchronized traffic signals that are adjusted in response to traffic conditions. From fiscal years 2003 through 2014, DOT provided about $570 million in funding for connected vehicle technologies. Funding for these efforts ranged from a low of $17 million in 2008 to a high of $84 million in 2011. These figures are not adjusted for inflation.

[10] For example, due to the sharing of data between vehicles, V2V technologies are capable of alerting drivers to potential collisions that are not visible to existing sensor-based technologies, such as a stopped vehicle blocked from view or a moving vehicle at a blind intersection. See GAO, *Intelligent Transportation Systems: Vehicle-to-Vehicle Technologies Expected to Offer Safety Benefits, but a Variety of Deployment Challenges Exist*, GAO-14-13 (Washington, D.C.: Nov. 1, 2013).

[11] 79 Fed. Reg. 49270 (Aug. 20, 2014).

[12] GAO-14-13.

[13] Along expressways linking cities in Japan, ITS Spots are installed approximately every 10-15 kilometers and on inner-city expressways, ITS Spots are installed every 4 kilometers. An in-vehicle V2I application collects, stores, and uploads (via ITS Spot road infrastructure) an anonymous travel and behavior record of the vehicle, which contains information about the time, position, speed, acceleration, and angle of the vehicle.

[14] According to DOT, DSRC technology is the designated communications technology for communications-based active safety-systems research.

[15] DSRC is used for safety-critical applications that cannot tolerate interruption; however, DOT has noted that other technologies (such as cellular or satellite, among others) may be used for non-safety-critical applications.

[16] Latency refers to the relative response time in communications between the originating and the responding application components (onboard unit and/or roadside unit and/or back office services) needed for the application to be effective.

[17] Radio-frequency spectrum is a natural resource that is used to provide an array of wireless communications services critical to the U.S. economy and a variety of government functions.

[18] In the United States, responsibility for spectrum management is divided between two agencies: FCC and the Department of Commerce's National Telecommunications and Information Administration (NTIA). FCC manages spectrum use for nonfederal users, including commercial, private, and state and local government users and NTIA manages spectrum for federal government users and acts for the President with respect to spectrum management issues. Historically, concern about interference or crowding among users has been a driving force in the management of spectrum. In order to minimize interference, FCC and NTIA have allocated particular bands of spectrum for specific uses, and provided users with a license or authorization to use a specific portion of spectrum. According to NTIA, the FCC will issue the licenses for the non-federal DSRC systems. DOT's spectrum use is authorized by NTIA. 75 Fed. Reg. 38387 (July 1, 2010).

[19] Amendment of Parts 2 and 90 of the Commission's Rules to Allocate the 5.850-5925 GHz Band to the Mobile Service for Dedicated Short Range Communications of Intelligent Transportation Services, Report and Order, 14 FCC Rcd 18221 (1999)(*FCC 99-305*).

[20] As new spectrum-dependent technologies and services are brought to market and government users develop new mission needs, the demand for spectrum continues to increase and additional capacity will be needed to accommodate future growth that cannot be addressed through more efficient use of wireless technologies. One driver of the increased demand for spectrum has been the significant growth in commercial wireless broadband services, smart phones, and tablet computers. See GAO, *Spectrum Management: FCC's Licensing Approach in the 11, 18, and 23 Gigahertz Bands Currently Supports Spectrum Availability and Efficiency*, GAO-13-78R (Washington, D.C.: November 20, 2012).

[21] In the Matter of Revision of Part 15 of the Commission's Rules to Permit Unlicensed National Information Infrastructure (U-NII) Devices in the 5 GHz Band, *28 FCC Rcd 1769, (2014).* The operator on an unlicensed device must accept whatever interference is received from the DSRC devices and must correct whatever interference is caused to DSRC devices. 47 C.F.R. § 15.5.

[22] When referring to the DSRC radio alone, the term roadside unit (RSU) is used.

[23] Currently, state and local traffic-management centers gather and process traffic data, such as information on accidents and congestion, and take steps to respond to conditions, such as notifying emergency personnel, adjusting traffic signals, and providing alerts to drivers via roadway signs.

[24] 79 Fed. Reg. 49270 (Aug. 20, 2014).

[25] DOT's ITS-JPO is responsible for research execution and initial technology transfer activities, such as field testing. FHWA is coordinating with the states and developing the materials to support V2I deployment, such as guides, tools, and best practices—and is working to ensure that deployed services are geographically interoperable and that deployed services are developed in accordance with federal regulations. The Volpe National Transportation Systems Center provides technical support to ITS-JPO and FHWA, and conducts some laboratory testing for V2I technologies, as well as estimating safety benefits of connected vehicle technologies.

[26] The report concluded that the transit safety applications have the potential to improve driver behavior and increase driver safety, but that improvements were needed to increase the accuracy of the warnings provided by the transit safety applications. For example, the report noted that the curve speed warning was provided early enough for the driver to take action only 57 percent of the time. U.S. Department of Transportation's Federal Highway Administration, *Independent Evaluation of the Transit Retrofit Package Safety Applications,* FHWA-JPO-14-175 (Washington, D.C.: February 2015).

[27] The primary members of the Connected Vehicle Pooled Fund Study include FHWA and transportation representatives from Virginia, California, Florida, Michigan, Minnesota, Maricopa County (Arizona), New Jersey, New York, Pennsylvania, Texas, Utah, Washington, and Wisconsin. Virginia serves as the lead state, with support from the University of Virginia.

[28] In addition, the University of Michigan's Mobility Transformation Center is launching a test facility to research automated vehicle technologies, including V2I.

[29] DOT has provided examples of the types of V2I applications that can be included in the pilot projects, and has noted that the applications used will be influenced by what local need the project is trying to address. For example, a pilot project may be located in a rural area with extreme weather and may include applications that improve access to weather-related warnings and that improve safety at highway crossings. Another project may be located in an urban area with poor air quality and thus include applications that address congestion, pedestrian safety, and vehicular emissions in the downtown area.

[30] As previously noted, unlike V2V, DOT is not considering mandating the deployment of V2I.

[31] For example, some of the tools FHWA is developing include a V2I Benefit Cost Analysis Tool; *V2I Planning Guide*; *Guide to V2I Cyber-Security*; *Guide to Licensing DSRC Roadside Units*; *Guide to V2I Communication Technology Selection*; V2I Message Lexicon (a list of allowable standard messages and formats for transmitted information for in-vehicle use); *Guide to Initial Deployments*; and Warrants for Deployment (a set of criteria which can be used to define the relative need for and appropriateness of a particular V2I application).

[32] AASHTO is a nonprofit association that serves as a liaison between state departments of transportation and the federal government.

[33] American Association of State Highway and Transportation Officials, *National Connected Vehicle Field Infrastructure Footprint Analysis* FHWA-JPO-14-125 (Washington, D.C.: June 2014).

[34] The Institute of Transportation Engineers is an international educational and scientific association of transportation professionals who are responsible for meeting mobility and safety needs.

[35] The Intelligent Transportation Society of America (ITS America) is a national organization dedicated to advancing the research, development, and deployment of Intelligent Transportation Systems (ITS) to improve the nation's surface transportation system. Founded in 1991, ITS America's membership includes more than 450 public agencies, private sector companies, and academic and research institutions.

[36] DOT officials noted that this will be the final version of initial guidance, and that FHWA intends to update this guidance over time.

[37] Equipping cars with V2V technologies should allow them to receive V2I messages from roadside infrastructure; however, it is possible that stand-alone, after-market safety devices could be purchased to equip existing vehicles.

[38] AASHTO cited research from the March 2009 Noblis report, *Footprint Analysis for IntelliDrive^{SM} V2V Applications, Intersection Safety Applications, and Tolled Facilities,* which found that twenty percent of intersections in the three largest metro areas accounted for 50 percent of the collisions. However, the study does not directly address but only infers the safety benefits of V2I-enabling signalized intersections.

[39] In a previous GAO report, we found that that the scarcity of spectrum in the United States is to some extent a result of the manner in which this resource has been allocated, managed, and used, rather than because of a physical scarcity of the resource. GAO, *Spectrum Management: Incentives, Opportunities, and Testing Needed to Enhance Spectrum*, GAO-13-7 (Washington, D.C.: November 2012).

[40] Pub. L. No. 112-96, §6406, 126 Stat. 156, 231.

[41] Traditional unlicensed equipment consists of low powered devices that operate in a limited geographic range, such as garage door openers and devices that offer wireless access to the Internet. They include Wi-Fi-enabled local area networks and fixed outdoor broadband transceivers used by wireless Internet service providers to connect devices to broadband networks.

[42] Spectrum sharing can be defined as the cooperative use of common spectrum that allows disparate missions to be achieved. In this way, multiple users agree to access the same spectrum at different times or locations, as well as negotiate other technical parameters, to avoid adversely interfering with one another. For sharing to occur, users and regulators must negotiate and resolve where (geographic sharing), when (sharing in time), and how (technical parameters) spectrum will be used. GAO-13-7.

[43] In the Matter of Revision of Part 15 of the Commission's Rules to Permit Unlicensed National Information Infrastructure (U-NII) Devices in the 5 GHz Band, Notice of Proposed Rulemaking, 28 FCC Rcd 1769 (2013).

[44] Two of 21 experts we interviewed did not provide a response to this question.

[45] In 2001, the Japanese government dedicated the 5.8GHz band to broadcast safety information using DSRC, and in 2011, dedicated the 700 MHz (760 MHz) band to support V2I technologies.

[46] In this case, latency refers to the relative response time in communications between the originating and the responding application components (onboard unit, roadside unit, or back office services) needed for the application to be effective.

[47] See Department of Commerce, NTIA. *Evaluation of the 5350-5470 MHz and 5850-5925 MHz Bands Pursuant to Section 6406(b) of the Middle Class Tax Relief and Job Creation Act of 2012* (Washington, D.C.: January 2013).

[48] DOT, *Fiscal Year 2016 Office of Secretary Congressional Budget Justification* (Washington, D.C.: February 2015).

[49] NTIA works with FCC via the Policy and Plans Steering Group, as well as the Policy and Plans Steering Group Spectrum Working group, to foster dialogue between agencies. In addition to the Steering Group, DOT has submitted information through NTIA to the FCC as part of the public docket process.

[50] IEEE members are engineers, scientists, and allied professionals whose technical interests are rooted in electrical and computer sciences, engineering, and related disciplines.

[51] The unlicensed devices would be expected to detect and vacate bands that are being used by DSRC operations at that time.

[52] The tests are developmental in nature and designed to demonstrate and measure a Cisco proposal in the presence of DSRC transmitters and to ensure the technology operates as

intended. Testing will begin in a laboratory setting and then advance to field tests, with the expectation that the initial round of feasibility testing will be completed by the end of 2015.

[53] The NCHRP is a research organization administered by the Transportation Research Board, and sponsored by members of AASHTO, in cooperation with FHWA. Individual projects are conducted by contractors with oversight provided by volunteer panels of expert stakeholders.

[54] See GAO. *High-Risk Series: An Update*, GAO-15-290 (Washington, D.C; February 2015). GAO maintains a program to focus attention on government operations that it identifies as high risk due to their greater vulnerabilities to fraud, waste, abuse, and mismanagement or the need for transformation to address economy, efficiency, or effectiveness challenges. Funding the nation's surface transportation system has been listed on our high-risk list since 2007.

[55] GAO, *Highway Trust Fund: All States Received More Funding Than They Contributed in Highway Taxes from 2005 to 2009,* GAO-11-918. (Washington. D.C: September 2011).

[56] The stakeholder was from a transportation research organization.

[57] Five of 21 experts we interviewed did not provide a response to this question.

[58] NCHRP, *Costs and Benefits of Public-Sector Deployment of Vehicle-to-Infrastructure Technologies, 03-101 (Washington, D.C.:* 2013).

[59] GAO, *Intelligent Transportation Systems: Improved DOT Collaboration and Communication Could Enhance the Use of Technology to Manage Congestion*, GAO-12-308 (Washington, D.C.: Mar. 19, 2012).

[60] DOT, *Lessons Learned from Safety Pilot and Other Connected Vehicle Test Programs*, (Ann Arbor, MI: May 30, 2014).

[61] For example, a deployment that supports V2I mobility or environmental applications may be eligible for funds through the Congestion Mitigation and Air Quality (CMAQ) Improvement Program, because these applications may provide the benefits of relieving traffic congestion, enhancing transit bus performance, and improving air quality.

[62] Center for Automotive Research and the Michigan Department of Transportation, *International Survey of Best Practices in Connected and Automated Vehicle Technologies*, (September 26, 2014).

[63] One of 21 experts we interviewed did not provide a response to this question.

[64] According to DOT, it is important to note that these are not design standards; for example, they do not specify specific products or designs to use.

[65] Before 2007, in the early development phase, DOT worked with National Institute of Standards and Technology on the framework for V2I standards.

[66] SAE International is a global association of more than 138,000 engineers and related technical experts in the aerospace, automotive, and commercial-vehicle industries. SAE International is recognized as the world's largest automotive and aerospace standards-setting body. According to SAE International, these standards are recognized as the foundation for safety, quality, and the effectiveness of products and services across the global mobility-engineering industry.

[67] According to the AASHTO Footprint Analysis, a mature connected vehicle environment by 2040 will include: (1) 80 percent (250,000) of traffic signal locations will be vehicle-to-infrastructure (V2I)-enabled; (2) up to 25,000 other roadside locations will be V2I-enabled; and (3) accurate, real-time, localized traveler information will be available on 90 percent or more of roadways.

[68] The Connected Vehicle Reference Implementation Architecture (CVRIA) is an initiative to define a multi-view system architecture to support development of full-scale connected vehicle deployments as well as to identify candidate interfaces for standardization.

[69] Some European nations have begun testing and deploying V2I efforts. For example, as previously noted, the Cooperative ITS Corridor will provide warning to drivers of upcoming roadwork and other obstacles via V2I technologies in three countries: Netherlands, Germany, and Austria.

[70] DOT, *An Approach to Communications Security for a Communications Data Delivery System for V2V/V2I Safety,* FHWA-JPO-11-130 (Washington. D.C: November 2011).

[71] One of 21 experts we interviewed did not provide a response to this question.

[72] This included 6 of 7 experts associated with a test bed, and 8 of 14 experts that were not associated with a test bed.

[73] One university researcher was certain that information could be stored and secured for a larger deployment.

[74] One of 21 experts we interviewed did not provide a response to this question.

[75] The request for information was for V2V; however, SCMS is expected to serve the same purpose for V2I efforts, in terms of managing security credentials.

[76] GAO-14-13. [77]GAO-14-13.

[78] One of 21 experts we interviewed did not provide a response to this question.

[79] One of 21 experts we interviewed did not provide a response to this question.

[80] According to DOT officials, the department plans to complete an assessment of V2I cyber security, and use the results to address the identified vulnerabilities for the V2I program. In addition, DOT will provide the results of this assessment to partners that will need to execute these mitigation measures. Currently, DOT is at the beginning stage of developing cyber-security materials for state and local DOT operating agencies to address the risk of legacy systems connecting to CV systems. DOT officials did not provide us with a completion date for these efforts.

[81] According to the Information and Privacy Commissioner of Ontario, Canada, privacy by design is an approach to protecting privacy by embedding it into the design specifications of technologies, business practices, and physical infrastructures. According to DOT, for V2I, this would create a security system technically designed so that it would be very difficult to use the system to track vehicles or individuals, or otherwise discover any personally identifiable information.

[82] However, according to DOT officials, automobile manufacturers are also evaluating how to combine data from public infrastructure and on-board sensors in vehicles to affect control functions such as braking, throttling, and potentially steering to mitigate driver inattention or confusion.

[83] GAO-14-13. The term "human factors" refers broadly to how humans' abilities, characteristics, and limitations interact with the design of the equipment they use and the environments in which they function.

[84] Two of 21 experts we interviewed did not provide a response to this question.

[85] DOT officials stated this was an ongoing effort and did not provide us with an estimated completion date.

[86] Four of 21 experts we interviewed did not provide a response to this question.

[87] Eight of the 21 experts we interviewed did not provide a response to this question.

[88] Eight of the 21 experts we interviewed did not provide a response to this question.

[89] Specifically, Japan estimates that congestion on the Metropolitan Expressway, located in Tokyo, was reduced from 56.2 kilometers/hour/day in 2003 to 2.8 kilometers/hour/day in

2007, with a respective rise in electronic toll usage from 6.1 percent to 73 percent of vehicles.

[90] Eight of the 21 experts we interviewed did not provide a response to this question.

[91] Transportation Research Board National Cooperative Highway Research Program, C*osts and Benefits of Public-Sector Deployment of Vehicle-to-Infrastructure Technologies*, NCHRP 03-101 (Reston, VA: 2013).

[92] Eight of the 21 experts we interviewed did not provide a response to this question.

[93] U.S. Department of Transportation, National Highway Traffic Safety Administration, *Frequency of Target Crashes for IntelliDrive Safety System,* DOT-HS-811-381 (Washington, D.C.: October 2010).

[94] U.S. Department of Transportation, Federal Highway Administration, *Crash Data Analyses for Vehicle-to-Infrastructure Communications for Safety Applications,* FHWAHRT-11-040 (McLean, VA: November 2012). This statistic is based on assumptions including full deployment and 100 percent effectiveness of applications.

[95] U.S. Department of Transportation, *AERIS: Applications for the Environment: Real-Time Information Synthesis Identification and Evaluation of Transformative Environmental Applications and Strategies Project, Initial Benefit-Cost Analysis*, (Washington, DC: September 30, 2012).

[96] According to DOT officials, reports with the results of these models are underway and unpublished; however, they did provide us with copies of the current versions of the reports.

[97] Transportation Research Board National Cooperative Highway Research Program, C*osts and Benefits of Public-Sector Deployment of Vehicle-to-Infrastructure Technologies*, NCHRP 03-101 (Reston, VA: 2013).

[98] For example, the study cited other studies that found that agencies using probe data from vehicles (for example, data from vehicles that provide insight into traffic or pavement conditions), could collect reliable data with a 5 to 10 percent penetration rate among vehicles. University of Michigan Transportation Research Institute, *Connected Vehicle Infrastructure Deployment Considerations: Lessons Learned from Safety Pilot and Other Connected Vehicle Test Programs*, (Ann Arbor, MI: May 30, 2014).

[99] Eight of the 21 experts we interviewed did not provide a response to the question about quantifying V2I benefits apart from V2V benefits.

[100] The NCHRP is a research organization administered by the Transportation Research Board and sponsored by members of AASHTO, in cooperation with FHWA. Individual projects are conducted by contractors with oversight provided by volunteer panels of expert stakeholders.

[101] We use both the NCHRP and the AASHTO reports to provide global statements about potential costs, but we will only be using cost figures from the AASHTO report as it contains the most recent information on potential, average estimates and is being used by DOT for the basis of cost estimation tools.

[102] During development of the "*National Connected Vehicle Field Infrastructure Footprint Analysis* (FHWA-JPO-14-125)," there was an effort to develop an estimate for the total cost of deploying V2I systems on a national level, using the anticipated deployment costs, locations and timelines. However, according to DOT officials, the uncertainties in the cost elements are so significant that it was not deemed appropriate to generate an estimated cost for deploying V2I systems on a national level.

[103] Nine experts and five stakeholders mentioned one or both of the reports' cost estimates.

[104] "Per site" refers to each specific location where a roadside unit is deployed. For example, a unit could be deployed at an intersection. The "per site" costs would be multiplied by the number of units required by the region.

[105] In a connected vehicle environment, there will also be costs for the on-board equipment. NHTSA conducted an assessment of preliminary V2V costs and estimated that the total cost per vehicle will be approximately $341 to $350 in 2020, decreasing to $209 to $227 in 2058.

[106] Traffic signal controllers provide information on the signal phase (green, yellow, or red) and the amount of time remaining until the light changes to the DSRC radio, which then broadcasts the information to the vehicle.

[107] As previously mentioned, backhaul is the closed network communication links between back offices (or TMCs) and field installations (such as traffic signal controllers).

[108] According to DOT officials, these costs are based on limited, research installations, and are not necessarily reflective of planned deployments. Furthermore, as agencies renovate or upgrade existing systems (such as periodic replacement of traffic signal systems), they are already installing components necessary to support V2I deployments, such as fiber optic backhaul for the modern traffic signal controllers.

[109] In its report, AASHTO assumes a cost of $50 per device to develop a more accurate estimate of the annual operating costs.

[110] 79 Fed. Reg. 49270 (Aug.20, 2014).

[111] Seven of the 21 experts we interviewed did not provide a response to this question.

[112] Seven of the 21 experts we interviewed did not provide a response to this question.

[113] Five of the 21 experts we interviewed did not provide a response to this question.

End Notes for Appendix I

[1] GAO, *Intelligent Transportation Systems: Vehicle-to Vehicle Technologies Expected to Offer Safety Benefits, but a Variety of Deployment Challenges Exist,* GAO-14-13 (Washington D.C.: November 2013)

[2] Connected vehicle technologies involve vehicle-to-vehicle (V2V) and vehicle-toinfrastructure (V2I) technologies. V2V is a system designed to transmit basic safety information between vehicles to facilitate warnings to drivers about potential collisions. V2I is defined as the wireless exchange of safety and operational data between vehicles and roadside infrastructure, intended primarily to avoid or mitigate motor vehicle crashes.

[3] There are two test beds in Michigan, the Safety Pilot in Ann Arbor, Michigan, and in Southeast Michigan, in Oakland County.

[4] AASHTO, *AASHTO National Connected Vehicle Field Infrastructure Footprint Analysis,* (Washington, D.C.: June 27, 2014) and National Cooperative Highway Research Program (NCHRP), *Cost and Benefits of Public-Sector Deployment of Vehicle-to-Infrastructure Technologies,* 03-101 (Washington. D.C.: 2013).

[5] However, it should be noted that these reports are limited in their scope because their findings are based on small, connected vehicle infrastructure deployments to date.

[6] Although there are V2I efforts underway in Europe, we focused on Japan due to its 20 years of experience in developing and deploying V2I technologies.

In: Vehicle-to-Infrastructure Technologies ISBN: 978-1-63485-111-4
Editor: Albert Buchanan

Chapter 2

2015 FHWA Vehicle to Infrastructure Deployment Guidance and Products*

U.S. Department of Transportation

Guidance

Introduction

This guidance is intended to assist Federal Highway Administration (FHWA) staff and transportation system owner/operators deploy Vehicle to Infrastructure (V2I) technology not only in terms of the Federal-aid Highway program requirements but also practices to help ensure interoperability and efficient & effective planning/procurement/operations. It is not intended to be an introduction to the Connected Vehicle (CV) technology; however, there are several quick reference web links to videos, slide presentations, and information modules related to the fundamental principles of connected vehicle technology in the appendix.

Deployment of V2I technologies is not mandated and is not coupled with the National Highway Traffic Safety Administration's (NHTSA) advance notice of proposed rulemaking for Vehicleto-Vehicle (V2V) communications.[1] The NHTSA rulemaking will not require State and local DOTs to deploy any connected vehicle infrastructure. However, it is important for the State and

* This is an edited, reformatted and augmented version of a guidance document (Draft version 9) issued by the U.S. Department of Transportation, September 29, 2014.

local agencies to understand a) what the decision could mean to them, b) what they need to know to prepare for an emerging connected vehicle environment, and c) what investments could be made in light of federal highway aid limitations to leverage a nationwide fleet of equipped vehicles in support of State and local policy and operational objectives. The guidance,

associated guides, toolkits, and products are to be a useful resource to help those considering V2I deployment and to leverage developments in V2V communications.

New and emerging connected vehicle technologies offer an opportunity to significantly enhance safety through communications between vehicles. There is also a potential to leverage this technology to improve mobility and further enhance safety though interaction with the road infrastructure. Although in many respects this technology resembles traditional ITS deployments, it is different in many ways:

- The technologies go beyond “connected” to be cooperative, allowing data and information from many sources to be fused in real-time;
- It thus requires a level of national interoperability and functionality not found in today’s ITS deployments;
- Because of its cooperative nature, it requires attention to security and privacy beyond today’s ITS deployments;
- The basic technologies that form connected vehicle systems are evolving at a dynamic pace.

As such, these technologies require a focused set of guidance.

This guidance was developed through cross modal cooperation at the Federal, State, and local levels. This guidance is intended to assist transportation agencies in making appropriate investment and implementation decisions when deploying connected vehicle systems. It does not negate or replace existing processes, but rather is intended to provide a clear statement of FHWA policy regarding such processes as planning, funding, siting, procuring, using, and providing access, among other critical actions.

In the event that there are inconsistencies between this guidance and existing laws and regulations, the existing laws and regulations take precedence.

As deployments come online and the technology matures this guidance will be assessed for relevancy, compatibility with real world deployment issues, and newly enacted transportation and communication statues and regulations. When needed this guidance will be updated and modified.

Background: The Connected Vehicle Environment

Connected Vehicle Environments (CVE) are systems comprised of hardware, software, and firmware that allow for the dynamic transfer of data between vehicles and between vehicles and the infrastructure including, at a minimum, Wireless Access in Vehicular Environment (WAVE) messages defined in Society of Automotive Engineers (SAE) J2735 that are broadcast on Dedicated Short Range Communications (DSRC). Moreover, the SAE J2735 DSRC standard is implemented for safety critical applications which cannot tolerate interruption or may be a threat to life.

However, non-safety-critical applications can use the DSRC protocol stack or other wireless media. For example, there are non-critical circumstances were communications with infrastructure and handheld devices that are implemented more effectively using multiple technologies—DSRC, cellular or Long-Term Evolution (LTE), publicly available Wi-Fi, satellite, wide or local area networks, or internet. Some points of connection may exchange data applications and may be primarily (or solely) enabled through DSRC; some applications may be primarily enabled through cellular, Wi-Fi, or other communications modes; and some applications may dynamically switch between communications modes in real-time. In many cases, the applications and/or radio will be capable of selecting the best communication mode to use at any time.

Most recently, the U.S. DOT has referred to Connected Vehicle as vehicle-to-vehicle communication technology and has been associated with vehicle telematics that connect vehicles to various information applications.[2] Moreover, the Basic Safety is the primary message set proposed to send data between vehicle and between vehicles and the infrastructure under the CVE. The CV can transmit and receive basic safety messages (BSMs) following the WAVE protocol, established in Standard IEEE 802.11p which uses the ITS band of 5.9 GHz (5.85 – 5.925 GHz). The content and format of a BSM are defined in Standard SAE J2735. The CVE uses the metric system and all applications should be developed and operate under it.

V2I components of CV Environments include: Roadside Units (RSUs), Signal Phase and Timing (SPaT) enabled traffic signal controllers, data links between V2I components and a traffic management center (TMC) or other back office, and any sensors or relays that link to or serve these components. An individual V2I deployment site could support more than one application, strategy, or impact area (i.e., safety, mobility, air quality, etc.). Generally, distribution of certificates through the CV security operations is not considered a safety application. CV technology also should consider pedestrians, bicyclists, and other possible nonmotorized users that may be within highways rights-of-way.

Although autonomous vehicles are not addressed in this guidance, this technology is not precluded by the FHWA V2I deployment guidance and may, in fact, be addressed by this guidance in the future. Appendix A contains a list of definitions, symbols, and abbreviations.

We seek public feedback on the proposed V21 components. Should any modifications be made?

V2I Deployment Policy Statement

Vehicle-to-infrastructure (V2I) services will take advantage of and build upon emerging vehicle-based technologies being deployed to support vehicle-to-vehicle (V2V) services. FHWA has not conducted a full benefit cost anlaysis of deployment of V2I services, but DOT believes the deployment may result in significant safety benefits and may also result in mobility and environmental benefits; and will be of significant interest to State, regional, and local transportation agencies.[3] Deployment of these services will be strongly encouraged by FHWA, but will be voluntary. FHWA will develop materials needed to support deployment (e.g., guides, tools, and best practices); ensure that deployed services are geographically interoperable; and ensure that deployed services are developed in accordance with the requirements in Part 940 of Title 23 within the Code of Federal Regulations (23 CFR 940) and other applicable regulations.

The following sections define key areas of deployment guidance.

Planning

CV technologies, including V2I applications, have the potential for a fundamental advancement in surface transportation safety and operations. While interactions between vehicles and infrastructure have been mostly passive and detached across the network to date, CV technology will allow them to work together, both actively and cooperatively. While we do not yet know the magnitude, this could provide opportunities for congestion reduction, safety improvements beyond V2V safety applications, as well as improved traveler services and information. To fully realize its potential, V2I site deployments, operations, data access, and exchanges (among other functions) will require even closer collaboration between transportation modes, public jurisdictions, and the private sector.

To this end, Metropolitan Planning Organizations (MPOs), Local Public Agencies (LPAs), transit operators, and States should begin considering V2I strategies in their long range planning. Topics of discussion could include: a general understanding of the system, which application options work well under what conditions, pros and cons of each option, capital costs and availability of funding, integration with the existing systems including traffic management and communication networks, long term impacts, cooperation and coordination across MPO boundaries and across State boundaries, staff needs, and the integration of these options into the existing Statewide or Regional ITS Architectures.

The National ITS Architecture provides a common framework to guide planning, defining, and integrating intelligent transportation systems at the Statewide and regional levels.[4] It is a mature product that reflects the contributions of a broad cross-section of the ITS community (transportation practitioners, systems engineers, system developers, technology specialists, consultants, etc.). The architecture defines:

- The functions (e.g., gather traffic information or request a route) that are required for ITS;
- The physical entities or subsystems where these functions reside (e.g., the field or the vehicle);
- The information flows and data flows that connect these functions and physical subsystems together into an integrated system.

The Connected Vehicle Reference Implementation Architecture (CVRIA) is an initiative that is defining the architecture views for connected vehicle technologies and identifying their key interfaces to analyze where standards may be optimal for interoperability or beneficial for operations. A standards development plan is in process to identify and prioritize candidate standards needed in support of connected vehicle implementation. The plan will consider adoption of existing industry standards from other areas, adaptation of existing standards, or development of new standards if existing standards are not viable. Ultimately, once the connected vehicle architecture views are mature, they will be incorporated into the National ITS Architecture.

To support planning for connected vehicle environments, the following work is ongoing:

- The FHWA is developing a planning guide to support analysis of investment options. A companion deployment guide is also under development to provide technical considerations for costing, siting, and installing connected vehicle technologies;

- Application requirements are emerging and can be found on the Connected Vehicle Reference Implementation Architecture website: www.its.dot.gov/arch/index.htm.

Federal-Aid Eligibility of V2I Equipment and Operations

A number of V2I activities are eligible for Federal-aid funding. However, specific program requirements should be met, and eligibility may need to be determined on a case-by-case basis.[5]

General Eligibility – Equipment, installation, preventive maintenance, and operational costs that support V2I applications and which are compatible with the basic CV standards for interoperability and security standards, are eligible for Federal-aid funding where eligibility for ITS investments has been previously established (also see "Transportation Systems Management & Operations: Operating Cost Eligibility Under the Federal-Aid Highway Program" – http://ops.fhwa.dot.gov/travelinfo/resources/ops guide.htm).

Expenses associated with maintaining CV systems are considered to be operating costs and are eligible for Federal-Aid funding. Examples of these costs include: system maintenance activities to ensure peak performance (such as preventive computer maintenance); replacement of defective or damaged computer components; or other traffic management system hardware (including street-side hardware). However, routine upkeep items that are not critical to maintaining continuous operation of the system (e.g. painting, grounds keeping, etc.) are not eligible for Federal-aid funding.

Marginal additional costs to purchase and install ITS equipment or communication technology capable of being upgraded or modified to operate in a deployed CV V2I environment at a future date are eligible for Federal Aid funding. In fact, it is recommended that as soon as CV technologies are available, such equipment should be the standard for deployment.

Procurement of spare "system-critical" parts (i.e., essential for the successful operation of the system) that are susceptible to failure, regardless of reason, are eligible if identified in a deployment maintenance and replacement plan (see reliability guidance).

Safety - V2I safety applications are eligible for HSIP funds if they address a State's Strategic Highway Safety Plan (SHSP) priority, are identified through a data driven process, and contribute to a reduction in fatalities and serious injuries. Additional information is available in the *HSIP MAP-21 Interim Eligibility Guidance.*[6]

Mobility – Costs that support V2I mobility and safety applications and which are compatible with the basic CV standards for interoperability and security may be eligible for NHPP and STP funds. NHPP-funded projects should be used for a facility located on the National Highway System (per 23 U.S.C. 119(c)) and should lead to advancement of the mobility, freight, condition and safety goals established at 23 U.S.C. 150(b) (per 23 U.S.C. 119(e)(2)). Just being one of the eligible projects listed at 23 U.S.C. 119(d) alone does not make the project NHPP eligible. Eligibility for funding under the Surface Transportation Program is determined by the provisions at 23 U.S.C. 133.

Congestion Mitigation and Air Quality (CMAQ) Improvement Program – Each CMAQ project should meet three basic criteria[7]: it should be a transportation project, it should generate an emissions reduction, and it should be located in or benefit a nonattainment or maintenance area.[8] All phases of eligible projects – not only construction- are eligible for CMAQ funding. This includes studies that are part of the project development pipeline (e.g., preliminary engineering) under NEPA. However, general studies that fall outside project development do not qualify for CMAQ funding. Additional information is available in the *CMAQ under MAP-21 Interim Program Guidance.*

FHWA determines the eligibility of all projects, except those related to transit. FTA determines the eligibility of transit projects. While the eligibility determination is not made jointly, the FHWA, FTA, and EPA field offices should establish and maintain a consultation and coordination process for timely review of CMAQ funding proposals, including those related to V2I.

ITS projects that support V2I applications, including equipment and installation costs, may be eligible for CMAQ funding provided they meet the basic eligibility requirements. CMAQ funds may also be used for operational assistance subject to the limitations as described in section VII.A.2 of the CMAQ Interim Program Guidance. Examples of V2I projects/applications, such as SPaT, eco-drive, Congested Intersection Adjustment, and traveler information systems, are eligible for CMAQ funds as they can be effective in relieving traffic congestion, enhancing transit bus performance, and improving air quality.

Transit projects that support V2I applications may be eligible for CMAQ funds. The FTA administers most transit projects. For such projects, after the FTA determines a project eligible, CMAQ funds will be transferred, or "flexed," from the FHWA to the FTA, and the project will be administered according to the appropriate FTA program requirements. Section VII.F.6 of the CMAQ Interim Program Guidance included a detail discussion of CMAQ eligible transit projects.

Although use for purely research deployments with test vehicles (e.g. ITS JPO's Safety Pilot Model Deployment) would not meet the eligibility test, deployments where 1) fully functioning equipped cars (either factory or aftermarket installations) will be interacting with the technology and impacting mobility and 2) may ultimately be incorporated in the wider deployment of the technology, would be eligible.

V2I Deployments and NEPA

V2I deployments are covered under 23 CFR 771.117(c)(21) as a Categorical Exclusion (CE) for certain intelligent transportation systems (ITS).[9] Since the deployments by definition have the primary function to improve the efficiency or safety of a surface transportation system, support the CV security system and enhance passenger convenience, they are covered by this section. Additionally these deployments will be typically within the existing operational right of way, conform to the criteria in section 23 CFR 771.117 (a), and are not likely to trigger the unusual circumstances provisions discussed in 23 CFR 771.117 (b).

Interoperability

To operate properly, V2I deployments should to be compatible with any CV security policies that may be developed, support the distribution and/or receipt and use of security certificates to the maximum extent possible, and protect privacy at the highest level appropriate to the performance of a Connected Vehicle (CV) Environment.

An operable CVE relies on the exchange of information, predominantly a Basic Safety Message (BSM) emanating from vehicles, but other messages as well.[10] Specifically, FHWA recommends that a BSM conform to SAE J2735 and should also be "signed" with a valid certificate in order to establish a trust relationship with those receiving and acting upon the message. We request comments on this proposed SAE J2735 conformation. Since vehicles and devices may interact with equipment from all over the Nation, the requirements for transmission and receipt of BSMs, including certificates, should be standard across the Nation and across vehicle makes and models; device makes and models; and applications when related to safety-critical messages. Similarly, V2I deployments should conform to these same requirements in order for successful exchange of information between vehicles, devices, and applications within the CV environment. Standard SAE J2735 establishes the current accepted message set for a connected vehicle environment, although this is subject to change as part of NHTSA's ongoing V2V rulemaking process.[11]

Although deployment of V2I components is voluntary, to the greatest extent possible, the functionality and utility of deployed applications will be consistent between transportation modes/regions and for deployed V2I equipment and/or applications to be eligible for Federal – aid funding and should comply with existing ITS and CV guidance/processes provided by U.S. DOT[12]. Upcoming reference manuals and reports will highlight where local choice is an option or where uniformity is required or offers a greater benefit.

Table 1 illustrates the types of messages, and the message handler, required to enable a variety of V2I applications. Note that the list of applications in Table 1 is illustrative and not exhaustive.[13] It is anticipated that industry will create additional applications that serve driver needs and agency requirements.

Evaluation

Due to the newness of CV - V2I technology it is important to determine the effectiveness in meeting an identified need, benefits/cost, and user satisfaction assessment of these early deployments and the value of the investment. Evaluations are critical to understanding not only the technology as a whole but the specific applications, how/when to deploy, and assist in the planning process. There are general ITS project resources that will be useful for this purpose; for example the ITS Evaluation Guidelines – ITS Evaluation Resource Guide found at www.its.dot.gov/evaluation/eguide resource.htm.

ITS Equipment Capability and Compatibility[14]

Early deployments of connected vehicle field infrastructure are likely to be installed alongside (or as part of) existing ITS equipment (e.g. Dynamic Message Signs, CCTV cameras, vehicle detections stations, etc.) and existing traffic signal controllers. One reason for this is that ITS deployments are already located in areas where V2I communications are likely to be most needed and beneficial. Also, these locations provide an opportunity to leverage existing power sources, cabinet space, and backhaul communications, which will minimize deployment costs. Installation of connected vehicle field infrastructure is conceptually no different than installation of other ITS equipment. The same considerations of siting, foundations, mounting points, power, physical accessibility and security, backhaul networks, and so forth that have become standardized and accepted in ITS practice, will be considerations in connected vehicle infrastructure deployments. In addition, new requirements may need to be considered for connected vehicle infrastructure deployments; such as, ensuring adequate line of sight for antennas and conducting mapping surveys of the surrounding roadway geometry.

In many cases, connected vehicle infrastructure will need to be integrated with existing field equipment in order to enable direct data communications between the infrastructure and vehicles. One example of this is where traffic signal controllers have new CV applications integrated into their functionality in order to broadcast signal phase and timing information directly to vehicles for safety and mobility applications. This type of integration of connected vehicle equipment with existing field equipment will require the use of standard interfaces and message sets that are currently being developed by FHWA and will be provided in V2I Reference Implementation documents (expected in late 2015).

In many ways, one can think of connected vehicle infrastructure as the next generation of ITS equipment being installed in the field with the potential to have a transformational impact on transportation operations and safety. As such, it is important to start considering V2I communications requirements and standards when new ITS equipment and traffic signal controllers are purchased and installed (procurement and installation guidance are described later in this document).

FHWA highly recommends that for any ITS equipment and traffic signal controllers, purchased in the future, and the deploying agency follow the systems engineering process and deploy the equipment in an environment that is CV ready. This guidance defines CV as a roadside installation that has the following characteristics:

- Reliable power supply;
- Two secure backhaul communications links, if required by implementing agency (one for ITS or traffic signal data and one for connected vehicle data);
- ITS equipment or controllers that are NTCIP standards compliant;

Table 1. V2I Messages by Application and Message Handler

	Position Correction Message Handler	Map Message Handler		BSM/ Approaching Vehicle Agregator				Traffic & Rail SignalMessage Handler			Local Weather and Road Surface Message Handler (Map Overlay)		Roadside Signage DMS Arbitrator	Local/Back Office User Interface
	Position Correction Message	Local Map Message	Traffic Signage Data	Approaching Vehicle Mess	BSM (V2I)	Vehicle Weather Data	Oversize Vehicle Detection	SPaT Messages (SRM, SSM)	Rail Crossing SPaT	Ped Crossing SPaT	Local Weather Message	Road Surface Condition Mess	Dynamic Roadside Sign Mess	Installation, Configuration, Maintenance
INFLO-Queue Warning (Q-WARN) (TME Based)	I	I		I	I	I							I	I
INFLO-Speed Harmonization (SPD-HARM) (TME Based)		I		I	I	I							I	I
Intelligent Traffic Signal System (ISIG)	I	I			I			I						I
Emergency Vehicle Priority	I	I			I			I						I
Transit Signal Priority	I	I			I			I						I
Freight Signal Priority	I	I			I			I						I
Pedestrian Mobility		I			I			I		I				I
Eco-Traffic Signal Timing;	I	I			I			I						I
Eco-Approach and Departure at Signalized Intersections;	I	I			I			I						I
Eco-Traffic Signal Priority; and	I	I			I			I						I
Eco-Driving - Connected Eco-Driving.	I	I			I			I						I
Dynamic Low Emissions Zones	I	I			I			I						I
Dynamic Eco-Lanes	I	I			I			I						I

	Position Correction Message Handler	Map Message Handler		BSM/ Approaching Vehicle Agregator				Traffic & Rail SignalMessage Handler			Local Weather and Road Surface Message Handler (Map Overlay)		Roadside Signage DMS Arbitrator	Local/Back Office User Interface
	Position Correction Message	Local Map Message	Traffic Signage Data	Approaching Vehicle Mess	BSM (V2I)	Vehicle Weather Data	Oversize Vehicle Detection	SPaT Messages (SRM, SSM)	Rail Crossing SPaT	Ped Crossing SPaT	Local Weather Message	Road Surface Condition Mess	Dynamic Roadside Sign Mess	Installation, Configuration, Maintenance
Enhanced Maintenance Decision Support System (MDSS).	I	I			I	I					I	I		I I
Information for Maintenance and Fleet Management Systems.	I	I			I	I					I	I		I
Weather-Responsive Traffic Management.		I	I	I	I						I	I	I	I
Motorist Advisories and Warnings.		I	I	I	I						I	I	I	I
Information for Freight Carriers.	I	I			I	I					I	I		I
Information and Routing Support for Emergency Responders.	I	I			I	I					I	I		I
Red-Light Violation Warning (RLVW)		I	I					I			I	I		I
Stop Sign Gap Assist (SSGA)		I	I	I	I								I	I
Curve Speed Warning (CSW)		I	I	I	I						I	I	I	I
Stop Sign Violation Warning (SSVW)		I	I								I	I		I
Railroad Crossing Violation Warning (RCVW)		I	I					I	I		I	I		I
Spot Weather Information Warning (SWIW)		I	I	I	I						I	I	I	I
Oversize Vehicle Warning (OVW)	I	I	I	I	I		I						I	I
Reduced Speed Zone Warning (RSZW)	I	I	I	I	I						I	I	I	

- Electronic map or geometric description of the surrounding area available in SAE J2735 compliant format;
- Roadside cabinet space sufficient to house an external processor (size depends on intended application) that may be installed in the future; and
- Considerations / plans regarding future mounting location(s) for a DSRC Roadside Unit (RSU) that will enable future V2I communications.

Connected vehicle infrastructure deployment standards and practices are still in development and will continue to be so until applications have been deployed and operating for several years. Practical field experience with connected vehicle system and applications deployment will continue to accumulate as new prototypes and model deployments are brought into the field. The Safety Pilot Model Deployment has significantly scaled up the field deployment of both infrastructure equipment and back office systems relative to prior test beds and demonstrations. In addition, U.S. DOT plans to initiate more connected vehicle pilot deployments in the coming years with an emphasis on V2I applications. Lessons learned and best practices from these pilot deployments will be invaluable guides to future deployments and will be the basis for updates to FHWA's V2I Reference Implementation and this V2I Deployment Guidance.

Hardware/Software Device Certification

It is the U.S. DOT's intent to enter a Cooperative Agreement with one or more facilities that will conduct qualification and certification testing for various connected devices (e.g., Vehicle Awareness Devices and Roadside Equipment), and applications used in large-scale connected vehicle deployment.[15] The Cooperative Agreement would identify an approach for a Connected Vehicle Certification Testing program. U.S. DOT has proposed a four (4) layer approach to connected vehicle devices and applications certification that includes Security Credential Management Systems (SCMS).

The four (4) layer approach consists of certifying a basic device to meet environmental and communications protocol requirements. Moreover, a basic device consists of hardware, software and radio components. The four (4) layers of certification are:

1. Environmental Abilities (e.g., temperature, vibration, weather);

2. Communication Protocol Abilities (e.g., the radio service interoperability for DSRC);
3. Interface Abilities (i.e., both the message syntax and contents are formatted properly);
4. Overall Application Abilities (i.e., verifies the system level function)

A basic device should be certified at layers one (1) Environmental Abilities and two (2) – Communication Protocol Abilities. However, an application should be certified in all four layers if it resides on a basic device. Therefore, this structure allows for multiple applications to be hosted on one device and use lower layer certifications (layers 1, 2, and 3). U.S. DOT will make available the certification facilities and SCMS certificates once they are finalized or receive the proper approvals.

Reliability

Typically, deployment of V2I Safety applications will augment existing devices, signs, or signals. However, given that the deployment is meant to improve safety or mitigate a design exception, operational reliability is paramount and should be incorporated in the design, procurement, and deployment.

V2I mobility applications will not necessarily backup existing schemes and have standalone functionality.

Reliability of deployed equipment should be incorporated into the system engineering process.[16] To ensure availability of the application, an equipment maintenance and replacement plan should be established holistically for all CV deployments. The idea of this concept and a resulting set of maintenance requirements build on the proven Systems Engineering approach. The systems engineering approach is recommended as the preferred method for developing ITS projects with FHWA's Rule 940.[17] (see http://ops.fhwa.dot.gov/docs/tmsmaintcptandplans/chapter3.htm).

Equipment and applications deployed on the NHS or using Federal-aid highway funds will be purchased from a qualified provider and will be certified through an industry approved process when available.

Use of Right of Way

Use of Right of Way (ROW) for V2I road-side unit (RSU) follows current regulations and funding eligibility. The State DOT should consider the engineering requirements, roadway safety, and sight distances when choosing the location of the V2I RSU in the right-of-way.

Installation of connected vehicle infrastructure within the ROW will be allowed as long as its use has a public benefit and does not impair the safety of the roadway. Private sector secondary use is permissible as long as it 1) does not interfere with or degrade the current and future primary mobility/safety applications; 2) the private application will be opt-in and the highway user will be able to disable at any time, at no cost to the user; and 3) a fair market value of the private application's use will be established.[18]

Private use arrangements are subject to 23 U.S.C. 156 and require the State to charge the fair market value for non-highway use or to obtain an exception from FHWA based on "a social, environmental, or economic purpose." (§ 156(b) and 23 CFR 710.403(d)). For Interstate right-of-way use an airspace agreement is necessary (23 CFR 710.405(d)). The Federal share of net income from the revenues obtained by a State from the use of the right-of-way should be used by the State for title 23 projects; preferably for deployment and operation of the V2I network.

For Federal purposes, secondary use messages are not considered advertising (e.g., displayed in a vehicle or on a personal device) under the Highway Beautification Act; however State law or policies could supersede this distinction.

Allowance of Private Sector Use[19]

FHWA supports the ability to maximize the possibility of private investment to leverage costs for deployment and operations. Public Owners/Operators of infrastructure communication systems (fiber optics, radio relay, etc.) can allow off-public right-of-way private sector use of the system for V2I communication as long as there is a public sector benefit. This should include for following:

- There is a public sector benefit of the V2I RSU;
- The private sector use does not interfere with or degrade the primary safety / mobility applications;
- The private sector application will be opt-in by highway users and highway users will be able to disable the application at any time;
- The roadway facility owner and/or Public Owners/Operators of the infrastructure system are not charged for any private sector use.

Public Owners/Operators of the infrastructure communication system may use, for example, a privately owned RSU to:

- Control traffic management signals and/or devices;
- Monitor roadway traffic performance;
- Distribute security certificates to vehicles, devices, or applications;
- Provide traveler information to vehicles through V2I communications.

The cost associated with installation and operation of private sector use or components would not be eligible for Federal-aid beyond the operation of the communication backhaul.

Design Considerations for Facilities

V2I applications may be utilized to mitigate safety and operational impacts that arise due to substandard geometric features of highways. This can be done at the network level or as part of a design exception. Therefore, the minimum design standards can be modified to take into account the safety and mobility benefits.

To operate in coordination with the national level, deployed equipment needs to be interoperable and coordinated with other modes of transportation (i.e. light and heavy duty vehicles, transit systems, railroad crossings, etc.). To the maximum extent possible, all devices and applications deployed by jurisdictions and modes should leverage this technology between the transportation entities and should not be standalone deployments.

Projects to construct or reconstruct highways should be designed in a manner that accommodates the concurrent or possible future installation of V2I roadside equipment.

Use of Existing Structures and Infrastructure

Installation of connected vehicle equipment on existing structures/ infrastructure will be allowed as long as its use has a public benefit and does not create potential safety issues. Private sector secondary use is permissible as long as it complies with the Uses of the Right of Way section contained elsewhere in this guidance. Designers should consider how CV technologies will affect pedestrians, bicyclists, and other nonmotorized users within the highway right-of-way and how CV technologies may affect access to transit services, and how CV technology can enhance livability.

Use on Public Sector Fleets (Including Incident Responder Vehicles)

Depending on specific program requirements, Federal-aid highway funds can be used to procure components that enable V2I applications that are installed on public sector vehicles. This includes all the components associated with both the collection of data and dissemination of that data from the vehicle to the infrastructure, as well as information dissemination from the infrastructure to the vehicle. Federal-aid funds can also be used to procure infrastructure-based components that enable V2I applications. This includes, and is not limited to, roadside equipment and components installed in maintenance garages and traffic/emergency operations centers. Instrumenting private contractor equipment performing maintenance/response or providing workzone safety capabilities can be federally funded as long as it provides a public benefit and the public sector retains ownership/control of the CV equipment.

To achieve interoperability, all of the above components will need to comply and be consistent with Connected Vehicle architecture and standards. Funds cannot be used for maintenance of any of the components described above unless identified in the equipment maintenance and replacement plan (see reliability guidance).[20]

The applications utilized on public fleets should comply with any guidance and recommendations that may be developed pertaining to demands on the driver's attention with respect to the collection of data (e.g., recording current road conditions) and presentation of information within the vehicle.

Through the system design process and the implementation of the applications, V2I messages should comply with Connected Vehicle Reference Implementation Architecture (CVRIA) and Basic Safety Message (BSM) standards.21 [22] Also, the messages should be prioritized in a manner that considers the immediacy/urgency of the message and its consequence (e.g., safety messages would take priority over mobility and environment messages). The applications deployed on public sector fleets should also be consistent and integrated with other CV deployments on the broader vehicle fleets (e.g., private sector vehicles).

Although the ability of first responders to preempt traffic signals is an important safety application, its unwarranted use could have serious mobility implications. If this application is deployed, it should be done in concert with guidance to those operators of vehicles with capability of preemption highlighting the negative impacts of misuse, when it should/should not be used, and disciplinary consequences of misuse.

Procurement Process

Quality procurements for the assets that support Connected Vehicle environments will enable consistent, secure, and interoperable implementtations. A generation of ITS deployment has provided many agencies with more mature practices for procurement. These practices can guide Connected Vehicle Environment procurements.

Evidence from hundreds of evaluation documents point to proactive stakeholder engagement as a determinant for deployment success. The key theme of stakeholder engagement is one that runs through the course of the entire procurement process, as the procurement will need to satisfy the needs of varied entities affected through the Connected Vehicle deployment.

There are several actions that are key to quality deployments:

- Identify what existing assets are candidates for modification to reduce the costs for delivery of power and communications to the deployment location. Whether in a rural or in an urban location, it is likely that there are various assets that are 'owned' by different entities. Gaining information on what assets are deployed, their operating capabilities, their supporting infrastructure (including mounting structures, power, and communications), and their provisions for maintaining the "legacy" assets will quickly focus the selection of candidate locations.
- Consult the ITS Costs Database for comparisons and ranges of unit and system costs. This database is fed from ITS program evaluation reports from various sectors on implementations of ITS. Much of the information generated in the ITS Costs Database originates with U.S. DOT model deployments and field operational tests. As CV test beds and pilot sites proliferate, unit cost information will be added to this database.
- Consult with the agencies that manage the statewide acquisition of information technology products to streamline the acquisition of connected vehicle assets (similar to procuring other ITS assets). Since most people outside the transportation community are not aware of Connected Vehicles, deployers will be expected to provide basic education and training on the technology and how it would benefit the public agencies. The FHWA is a resource for providing the basic information, and may be available to provide direct support for engaging with the IT investment authorities.
- Apply the specifications that are produced by U.S. DOT as the basis for amending their own existing Qualified Products Lists. Vendors and solution developers are most effective when there is less variation to the specifications toward which they are building. Establishing a

State or local Qualified Products Lists introduces a level of rigor that better secures the interoperability among agencies working in the same geographic area, and locations far from the implementation considered by the agency.

- Modify existing guides for project development to include connected vehicle assets as part of the project cost estimation. Amend the project development guide sections for ITS assets to feature the Connected Vehicle assets. Allow for contingencies consistent with other ITS roadside assets like dynamic message signs. Project cost estimation should consider the costs for all of the systems engineering components articulated in various guides available through FHWA. One such resource is Systems Engineering for Intelligent Transportation Systems available at http://ops.fhwa. dot.gov/ publications/seitsguide/index.htm. The FHWA will be developing new reference materials for project cost estimations.
- Apply the Connected Vehicle Footprint Analysis produced by AASHTO to consider the typical profile and plan view drawings that could be applied for developing implementation plans. The plans will need to reflect the assessments made previously on the existing assets and supporting infrastructure. The design guides that exist among the State DOT's should be modified to incorporate elements of the AASHTO examples to standardize drawing plans.
- Establish test procedures and systems acceptance methodologies that minimize the risk to the agency. Ensure that during systems acceptance that data produced from the locations are not the sole possession of the Design-Build contractor, if one is used, but that the data can be validated by the agency working through an independent party.
- Coordinate with the agency's asset management unit to maintain accurate records of when items were deployed. This establishes a baseline for determining the lifecycle of the units. This will support the efforts by the agency to develop appropriate funding plans to sustain the operation of the system over time.
- Establish an operations and maintenance plan for the deployed system. Similar to the initiation of the Connected Vehicle Environment implementation, the choices are to contract with design/build entities or to work with a turn-key provider. There may be advantages to consider with the type of operations and maintenance plan in delivering: staff with appropriate technical

expertise; responsiveness in repair once the environment is in an operational state; and, flexibility in obtaining support services through a number of providers.

Legacy Systems/Devices

Legacy systems and devices may be owned and operated by public agencies, private companies, or the general public. This guidance addresses those systems and devices that are under the jurisdiction of public agencies. The principles apply to other legacy systems and devices and may be applied through other agreements, coalitions, and similar cooperative efforts.

Legacy systems or devices critical to the function of active safety applications should be retrofitted or replaced in order for the safety application to continue to do its job. Other legacy systems and devices may be augmented by V2I safety applications that may be intended to improve safety or mitigate a design exception. Similarly, legacy systems and devices may support or be supplemented by other V2I applications. In these instances, retrofitting or replacement of legacy systems or devices may not be required in order for the V2I applications to provide their services. However, support for V2I and other connected vehicle applications and functions should be considered as these legacy systems and devices are upgraded and replaced.

The system engineering process should be used in establishing equipment maintenance and replacement plans for legacy systems. For replacement systems using Federal-aid funding, the system engineering process should be used per requirements in Part 940 of Title 23 of the Code of Federal Regulations (23 CFR 940). Among the issues to be considered in the system engineering analysis are the functionality necessary to support V2I application requirements, the number of devices affected, the useful life of equipment and life-cycle costs, and processes for replacing equipment.

Communication Technology

The selection of the V2I communications technology will be based on a systems engineering analysis and will be consistent with application interoperability across the Nation. Use should comply with established requirements for non-interference.[23]

Although DSRC is the anticipated communication technology for V2V safety applications, there could be other communication methods available or better suited for V2I mobility applications. The important factors when considering alternatives to DSRC are: capability of V2V On-board Units

(OBUs), national interoperability, certified application support, and the technology's attributes meeting the needs of the application/installation.

Deployers will be expected to reference the analyses/white papers/reference guides that FHWA is developing (see Products/Tools).

Dedicated Short Range Communications (DSRC) Service Licensing

On December 17, 2003, the Federal Communications Commission adopted a Report and Order establishing licensing and service rules for the Dedicated Short Range Communications (DSRC) Service, in the Intelligent Transportation Systems (ITS) Radio Service in the 5.850-5.925 GHz band (5.9 GHz band).[24] [25] [26] The DSRC service involves vehicle-to-vehicle and vehicle-toinfrastructure communication to protect and enhance the safety of the traveling public. Also, the band is also eligible for use by non-public safety entities and for private commercial operations.

However, the safety applications always have primary status over non-safety applications. Roadside infrastructure is licensed to both public safety and non-public safety entities pursuant to 47 CFR Part 90, while on-board units are licensed by rule (i.e., no individual license is required) under 47 CFR Part 95.

Therefore, the National Telecommunication and Information administration authorizes the use of this frequency spectrum for governmental entities while the Federal Communications Commission issues licenses for the private sector which is based on each applicant's area-ofoperation (i.e., by County, State, multi-State, or Nationwide). Those applicants who are approved will each be granted a non-exclusive license for the geographic-area requested. Given the likeliness of cross border and/or overlapping geographic-area issues, applications, and potential efficiencies, it is stressed that a high level of jurisdictional cooperation and coordination across these boundaries will be necessary.

Besides States/local governmental entities and non-public safety entities, there is a wider-group that can hold an authorization to operate Roadside units (RSUs) in the DSRC 5.9 GHz band; specifically entities meeting requirements of 47 CFR 90.33 and 90.35 of the FCC Rules. Entities engaged in the operation of a commercial activity can receive authorizations to provide commercial mobile radio service or to operate stations necessary for transmission of communications (i.e., toll collection, and parking guidance). Further it should be noted that the licensing requirement only applies to equipment used for transmission; equipment designed and operated only to receive do not require siting licenses.

Although safety communication has precedence, identification and removal of interfering non-safety signals could be problematic, particularly if the interfering deployment was licensed and sited first. FHWA recommends that once an application deployment is identified in the planning process, site licensing should be undertaken. To this end, a guide to DSRC licensing is being developed for use by public agencies.

Data Connection and Latency

The data connection and latency are key factors in the CV implementation to improve mobility and further enhance safety through interaction with the road infrastructure. DSRC can communicate directly between vehicles and infrastructure with low latency but it is a line-ofsight technology and has limited range. Therefore, the key infrastructural features and aspects of deployment may be fundamentally different for the same application using different communications technology (e.g., solely DSRC, DSRC and cellular, Wi-Fi, satellite, Internet). In many cases, the applications and/or radio will be capable of selecting the best communication mode to use at any time.

However, a reliable data connection and low latency should be maintained to ensure a safe and a reliable CV environment. The reliable data connection should be highly available based on three key principles: elimination of single point of failure, reliable crossover, and detection of failures in real-time. While, the low latency will be based on the communication media; however, it should be measured from end-to-end or between the originating and the responding application.

Connected Vehicle Privacy Principles

An operable Connected Vehicle Environment relies on the exchange of information between vehicles (V2V) and between vehicles and infrastructure (V2I). V2V applications are enabled through broadcast and receipt of a Basic Safety Message (BSM) by vehicles. The transmission of data to and from a particular vehicle gives rise to individual privacy concerns.

The U.S. DOT takes privacy very seriously. We are committed to supporting deployment of a CV environment that both protects personal privacy and promotes this important safety, mobility and environmental technology. For this reason, the U.S. DOT has worked closely with its partners to develop a technical approach to CV communications that helps protect individual privacy. CV equipment in motor vehicles transmits the generic information used by V2V and V2I applications in a very limited geographical range. These generic BSMs do not identify specific drivers or vehicles.

Additionally, the U.S. DOT and its partners incorporated into the design of the CV environment multiple mechanisms specifically to minimize risks to individual privacy, including a Certificate Management Security System (SCMS) (discussed below).

Also, the U.S. DOT has emphasized that deployment of technologies and applications in the connected vehicles environment should be guided by the Fair Information Privacy Principles (FIPPs), which are based on tenets of the Federal Privacy Act of 1974 and mirrored in the laws of many U.S. States. These include:

- Transparency- Ensure that consumers have information about the data being collected and transmitted by the V2I system and how that data will be used;
- Individual Participation and Redress- Ensure that consumers have a reasonable opportunity to make informed decisions about the collection, use, and disclosure of their PII or other data that may be used to identify them directly or indirectly, and reasonable access to their PII and the opportunity to have their PII corrected, amended, or deleted, as appropriate;
- Purpose Specification: Make clear the purpose(s) for which the V2I system collects, uses, maintains, or disseminates specific data elements, such as basic safety functions as well as potential mobility, environmental, or commercial applications;
- Data Minimization: Explain why the data collection isn't excessive and how long data will be retained;
- Use Limitation: Assure consumers that the collected data will not be used for purposes incompatible with the specified purposes;
- Data Quality and Integrity: Explain how the V2I system will assure data quality and integrity throughout the data lifecycle and in all associated uses;
- Security: Explain what physical, technical, and procedural measures system administrators will take to protect collected data;
- Accountability and Auditing: Explain how the V2I system will ensure that the privacy controls are executed.

FHWA will work to identify and disseminate data privacy best practices as applications and the V2I system develops.

Connected Vehicle Security

The Security Certificate Management System (SCMS) is a critical component of the CV environment designed to protect the security and privacy of the BSM data exchanged by vehicles and between vehicles and infrastructure. As currently envisioned, the CV SCMS will be a complex Public Key Infrastructure (PKI) that creates, manages, stores, distributes, and revokes digital certificates that accompany and validate each BSM. It is this PKI that will allows users of the CV environment's unsecure public network to securely and privately exchange BSM data through the use of a public and private cryptographic key pair that is obtained and shared through a trusted authority.

A PKI typically consists of the following elements:

- A certificate authority that issues and verifies digital certificates. A certificate includes the public key or information about the public key;
- A registration authority that acts as the verifier for the certificate authority before a digital certificate is issued to a requestor;
- One or more directories where the certificates (with their public keys) are held;
- A certificate distribution and management system which includes the communications system and its organizational and operational elements.

The U.S. DOT has determined that the PKI for the connected vehicles environment should be organized such that no single organization within the PKI holds enough information about a participant to link a BSM transmitted by a vehicle to a specific driver or identified vehicle.

Data Access

In general, Federal law does not assign ownership, access, and use limitations to broadcast data. As a result, the U.S. DOT and FHWA do not currently have a specific policy assigning data ownership or limiting access to BSM data.

There is a need to balance access to data for use in developing mobility, environmental, and other applications (including potential private sector applications) with the consumer's concerns regarding privacy. In order to address consumer concerns about privacy and enhance consumer acceptance, V2I applications should contain sufficient controls to mitigate potential privacy and security risks appropriately. Two critical controls directly relevant to data access are transparency and consent. Consistent with DOT's Connected Vehicle Privacy Principles, public and private sector application developers should ensure that consumers understand and, for opt-in applications, consent to the collection of their data.

The FHWA is evaluating these issues and will develop a white paper to explore such topics as the scope of data generated in the V2I system, the expected uses of such data, the public sector's need to access this data for safety, mobility, and environmental applications, and the private sector's role regarding the use and collection of V2I data.

Manual on Uniform Traffic Control Devices

V2I applications providing traffic control information to drivers should be consistent with the Manual on Uniform Traffic Devices (MUTCD).[27] It is not expected that every roadside sign, signal, or pavement marking will be mirrored within the vehicle. If they are, generally these displays will be implemented by the OEMs. The information the on-board unit (OBU) receives should be sufficient for it to generate the appropriate sign/symbols or convey that information in a manner consistent with the MUTCD. In addition, all information conveyed to the driver should comply with and cannot contradict information conveyed by the signs, signals and markings on and along the road (as defined by the MUTCD). In-vehicle systems should also convey priority captured by signs, signals and markings (e.g., regulatory signs take priority over warning signs).

Using Public-Private Partnerships (P3s)

To the extent possible P3s and other commercial relationships should be considered for deployment. However, such arrangements should assure that the "commercial" applications protect the public interest, do not compromise the safety and mobility objectives, provide for safe maintenance practices, and hold the jurisdiction harmless due to lack of both public and private services. These arrangements should be established through P3 agreements that guarantee the precedence of the V2I safety and mobility applications, including the ability to terminate the relationship in the event that the private sector use should degrade the V2V or V2I environment. The public owner/operator of the V2I infrastructure should also ensure that the arrangement produces a net life-cycle benefit to the public and value for the transportation agency. Additionally, such deployments should be compatible with the CV security system, support the distribution and receipt and use of certificates to the maximum extent possible, and uphold the privacy principles of Connected Vehicle (CV) environment.

Current FHWA policy allows State Departments of Transportation (DOTs) to use Federal-aid funds in innovative long-term contracts with private developers under certain conditions. In such agreements, the State grants exclusive rights (a "concession") to a developer (concessionaire) who assumes responsibility for the highway's construction, operations, and upkeep. In the case described here, the developer would take responsibility for the deployment and operation of the V2I system.

P3 concession contracts often allow the concessionaire to collect and retain revenue from tolls, but in this instance tolls would not be applicable. The “business model” of CV V2I P3 is different in the sense that, instead of the private entity receiving payment from the highway user for access to a toll lane or facility, revenue to the private entity would most likely come from use of the communications channel to generate fee-for-service opportunities or advertising revenue.

Additional information on transportation P3s can be found at http://www.fhwa.dot. gov/ipd/p3/index.htm.

CONCLUSION

It is recognized that since this initial guidance was developed concurrently with a period of rapid maturity of the Connected Vehicle technology, it will need to be continually assessed for relevancy, compatibility with real world deployment issues, the evolution of the CV technology and security, and new transportation and communication statues and regulations. As early deployments come online or jurisdictions take on planning activities, it will be important for lessons learned, identified gaps, or innovative approaches be acknowledged, shared, and incorporated into updates to this guidance and its associated guides and toolkits.

Products/Tools

To support the V2I Deployment Guidance a number of guides, best practices compilations, and toolkits are being developed. Below is a list of major efforts underway. Although all of these may not be completed and/or in their final form at time of guidance issuance, they are scheduled to be delivered in 2015. Much like the guidance it is anticipated that it will need to be continually assessed, updated, and supplemented to stay current with the CV state of art.

System Engineering Process for Vehicle to Infrastructure

Augment the Systems Engineering (SE) for Intelligent Transportation Systems manual and provide guidance for professionals involved in developing systems engineering documents covering the evaluation, selection and implementation of Connected Vehicle (CV) V2I technology.

V2I Benefit Cost Analysis Tool

Tool to assist adoption of an investment strategy that dedicates funding to the capital and ongoing operational costs for connected transportation.

V2I Planning Guide

Provide planning staff with an increased awareness of the benefits and opportunities for deploying these technologies by State DOTs and MPOs that guidance on inclusion of the technology in long range plans and project selection process.

Guide to V2I Cyber-Security

Provides deployers with: (1) an analysis of extensibility of security and trust system to additional points of connection, including V2I, devices, backhaul, and others; (2) an analysis of additional risks from extensibility and cyber security; and (3) an analysis of the potential impacts to the existing transportation system/network. It will also provide definitions of the organizational functions and processes for operating the security function, along with cost models for operations and maintenance.

Guide to Licensing DSRC Roadside Units

A guide to requirements that a transportation owner/operator will use to navigate the process of licensing, be they in a position to develop or manage an outsourcing contract or understand / deal with private sector commercial deployments.

Guide to V2I Communication Technology Selection

A description of the technology options available and why certain options may be more appropriate for some applications than others.

V2I Message Lexicon

A list of allowable standard messages and formats for transmitted information for In-Vehicle use. Although the OEMs will control the message/warning type provided to the user, what type of information and its form the RSU sends needs to be standardized.

Guide to Initial Deployments

A guide to transportation system owner/operators which are on a path from no V2I deployment to a build out of the various scenarios. This would include prioritization methodology, staged deployment applications, co-location with existing ITS infrastructure, legacy equipment, utility based on V2V market penetration, etc.

Warrants for Deployment

A set of criteria which can be used to define the relative need for and appropriateness of a particular V2I application.

Appendix A

This section has quick reference web links to videos, slide presentations, and information modules related to the fundamental principles of connected vehicle technology. These videos are described as Talking Technology & Transportation (T3) webinars and sponsored by the Intelligent Transportation Systems (ITS) Professional Capacity Building Program (PCB) under the Department of Transportation (DOT) ITS Joint Program Office (JPO). This section is divided into two (2) parts: Connected Vehicle Fundamental Concept Videos and Connected Vehicle Information Modules.

Connected Vehicle Fundamental Concept Videos

Connected Vehicle Basics

Summary of Video: The webinar features two speakers (Valerie Briggs & Brian Cronin) from the U.S. DOT who introduce the Connected Vehicle Research Program, explain the concept, describe the technology, and explain the benefits and challenges of connected vehicles. Also, the webinar discusses how connected vehicles affect safety, mobility, and the environment.

Video: http://www.pcb.its.dot.gov/t3/s140424 cv basics.asp

Presentation 1: http://www.pcb.its.dot.gov/t3/s140424/s140424 cv basics presentation lister.pdf

Presentation 2: http://www.pcb.its.dot.gov/t3/s140424/s140424 cv basics presentation briggs cronin.pdf

Question/Answer Transcript: http://www.pcb.its.dot.gov/t3/s140424/s140424 cv basics qa.asp

Vehicle-to-Vehicle Communication: A New Generation of Driver Assistance and Safety Summary of Video: ITS-JPO has developed an information video that demonstrates the following technologies: Dedicated Short Range Communications (DSRC), Vehicle to Vehicle (V2V) communication system, Blind Sport Warning (BSW), Forward Collision Warning (FCW), Emergency Electronic Brake Light (EEBL), Intersection Movement Assist (IMA), and Do Not Pass Warning (DNPW).

Video: http://www.its.dot.gov/library/media/v2v video.htm

NOTE: Immediately after this video, there are several more videos (10+) available which explain various V2V applications (or scenarios), CV testbed, CV Safety Pilot, Dedicated Short Radio Communication (DSRC), and so more.

Connected Vehicle Information Modules

Connected Vehicles – Module 13

Summary of Video: This module describes the background, current activities, and future direction of the connected vehicle initiative. The module examines the anticipated roles and responsibilities of the principal participants; the major technologies and systems development efforts; the range of expected applications of the connected vehicle system; the potential institutional, policy, legal, and funding challenges facing the initiative; and the expected development and deployment timeline for a connected vehicle environment.

Link: http://www.pcb.its.dot.gov/eprimer/module13.aspx

APPENDIX B

Definitions

Term	Description
Aftermarket Safety Device (ASD)	A connected device in a vehicle that operates while the vehicle is mobile, but which is not connected to the data bus of the vehicle.
Backhaul	The closed network communication links between a Traffic Management Center (or other back offices), links between TMCs, and field installations (such as traffic signal controllers, traffic cameras, and other sensors). This could also include the link between the Security Credential Management System and roadside distribution device.
Basic Safety Message (BSM)	The core data set transmitted by the connected vehicle (vehicle size, position, speed, heading acceleration, brake system status) and transmitted approximately 10x per second. A secondary set is available depending upon events (e.g., ABS activated) and contains a variable set of data elements drawn from many optional data elements (availability by vehicle model varies). This would be transmitted less frequently. The BSM is tailored for low latency, localized broadcast required by V2V safety applications but can be used with many other types of applications.
Connected Device	Any device used to transmit to or receive messages from another device. A connected device can be sub-categorized as an OBE, ASD, VAD, or RSE. In many cases the connected device will be a DSRC device, but other types of communications can and are expected to be supported.
Connected Vehicle (CV)	A vehicle containing an OBU or ASD. Note that vehicles may alternatively include a Vehicle Awareness Device (VAD), which transmits the BSM but does not received broadcasts from other devices and cannot directly support vehicle-based applications.
Connected Vehicle Reference Implementation Architecture (CVRIA)	A set of system architecture views that describe the functions, physical and logical interfaces, enterprise/institutional relationships, and communications protocol dependencies within the connected vehicle environment. The CVRIA defines functionality and information exchanges needed to provide connected vehicle applications.
Dedicated Short Range Communications (DSRC)	DSRC is a technology for the transmission of information between multiple vehicles (V2V) and between vehicles and the transportation infrastructure (V2I) using wireless technologies.
Intelligent Transportation Systems (ITS)	Systems that apply data processing and data communications to surface transportation, to increase safety and efficiency. ITS systems will often integrate components and users from many domains, both public and private.
Interoperability	The ability of two or more systems or components to exchange information and to use the information that has been exchanged. The dependence of the CV Environment on successful exchange of data between independent components results in a requirement that all V2I deployments.
Latency	A measure of time delay experienced in a system, the precise definition of which depends on the system and the time being measured. For a data element in this context, latency is the time difference between the time that data value is acquired by the source and the time the message is transmitted.

Term	Description
NTCIP	The National Transportation Communications for Intelligent Transportation System Protocol (NTCIP) is a family of standards designed to achieve interoperability and interchangeability between computers and electronic traffic control equipment from different manufacturers.
On-Board Equipment (OBE)	This term refers to the compliment of equipment located in the vehicle for the purpose of supporting the vehicle side of the applications. It is likely to include the DSRC radios, other radio equipment, message processing, driver interface, and other applications to support the use cases described herein. It is also referred to as the Vehicle ITS Station. When referring to the DSRC radio alone, the correct term is OBU (see below).
On-Board Unit (OBU)	A vehicle mounted device used to transmit and receive a variety of message traffic to and from other connected devices (other OBUs and RSUs). Among the message types and applications supported by this device are vehicle safety messages, a primary subject of this standard, used to exchange information on each vehicle's dynamic movements for coordination and safety.
Original Equipment Manufacturer (OEM)	An original equipment manufacturer refers to the entity that originally manufactures and item that may be branded and sold by others. In the Connected Vehicle Environment, it is commonly used to refer to automobile manufacturers.
Public-Private Partnerships (P3)	Public-private partnerships (P3s) are contractual agreements formed between a public agency and a private sector entity that allow for greater private sector participation in the delivery and financing of transportation projects.
Roadside Equipment (RSE)	Term used to describe the compliment of equipment to be located at the roadside; the RSE will prepare and transmit messages to the vehicles and receive messages from the vehicles for the purpose of supporting the V2I applications. This is intended to include the DSRC radio, traffic signal controller where appropriate, interface to the backhaul communications network necessary to support the applications, and support such functions as data security, encryption, buffering, and message processing. It may also be referred to as the roadside ITS station. When speaking of the DSRC radio alone, the correct term is RSU (see below).
Roadside unit (RSU)	A connected device that is only allowed to operate from a fixed position (which may in fact be a permanent installation or from temporary equipment brought on-site for a period of time associated with an incident, road construction, or other event). Some RSEs may have connectivity to other nodes or the Internet.
Signal Phase and Timing (SPaT)	In the context of this standard, SPaT is a message type that describes the current state of a signal system and its phases and relates this to the specific lanes (and therefore to maneuvers and approaches) in the intersection.
Security Certificate Management System (SCMS)	A public key infrastructure approach to security involving the management of digital certificates that are used to sign and authenticate messages among legitimate but unknown vehicles and/or equipment and/or other points of connection.
Systems Engineering	An interdisciplinary practice which focuses on how to design and manage complex projects/deployments over their life cycles. It ensures that all likely aspects of a system are considered and integrated into a whole.

Definitions (Continued)

Term	Description
Vehicle	A self-propelled transport device, along with any attachments (e.g., trailers), that is a legal user of the transportation network.
V2I Reference Implementation	An interface system that supports the collection, integration, and dissemination of data between infrastructure and vehicles to enable integrated, interoperable V2I safety, mobility, and environmental applications.
V2V	Short for vehicle-to-vehicle communications: a system designed to transmit basic safety information between vehicles to facilitate warnings to drivers concerning impending crashes.
V2I	Short for vehicle-to-infrastructure communications: a system designed to transmit information between vehicles and the road infrastructure to enable a variety of safety, mobility, and environmental applications.

Symbols and Abbreviations

Term	Meaning
BSM	Basic Safety Message
CV	Connected Vehicle
DSRC	Dedicated Short Range Communications
IEEE	Institute of Electrical and Electronics Engineers
ITS	Intelligent Transportation Systems
MUTCD	Manual on Uniform Traffic Control Devices
OBE	On-Board Equipment (synonym for On-Board Unit)
OBU	On-Board Unit
PTV	Public Transport Vehicle
RF	Radio Frequency
RSE	Roadside Equipment (synonym for Roadside Unit)
RSU	Roadside Unit
SAE	Society of Automotive Engineers International
SPaT	Signal Phase and Timing
TMC	Traffic Management Center
V2I	Vehicle to Infrastructure
V2V	Vehicle to Vehicle

End Notes

[1] U.S. Department of Transportation Issues Advance Notice of Proposed Rulemaking to Begin Implementation of Vehicle-to-Vehicle Communications Technology (NHTSA 34-14 issued August 18, 2014); http://www.nhtsa.gov/About+NHTSA/Press+Releases/NHTSA-issues-advanced-notice-of-proposed-rulemaking-onV2V-communications.

[2] DOT has long used the term "connected vehicle" to refer to the vehicle-to-vehicle communication technology that supports crash avoidance applications. However, the term has expanded to associate vehicle telematics that connect vehicles to various information applications through other forms of communications. Refer to Harding, J., Powell, eta, Vehicle-to-vehicle communication: Readiness of V2V technology for application; Report No. DOT HS 812 014, issued August 2014.

[3] Refer to Harding, J., Powell, eta, Vehicle-to-vehicle communication: Readiness of V2V technology for application; Report No. DOT HS 812 014, issued August 2014.

[4] Regional ITS Architecture Guidance Document Version 2.0, July 2006: On January 8, 2001, the US Department of Transportation published the FHWA Final Rule and FTA Policy, which implement section 5206(e) of the Transportation Equity Act for the 21st Century (TEA-21). TEA-21 required ITS projects funded through the highway trust fund to conform to the National ITS Architecture and applicable standards. http://ops.fhwa.dot.gov/publications/regitsarchguide/.

[5] Refer to FHWA Operations Regulation and Policy http://www.ops.fhwa.dot.gov/opsregulationpolicy.htm and Transportation Systems Management & Operations: Operating Cost Eligibility Under the Federal-Aid Highway Program. Reference http://www.ops.fhwa.dot.gov/travelinfo/resources/ops_guide.htm#fed

[6] http://www.fhwa.dot.gov/map21/guidance/guidehsip.cfm.

[7] http://www.fhwa.dot.gov/environment/air_quality/cmaq/

[8] http://www.fhwa.dot.gov/environment/air_quality/cmaq/reference/intelligent_transportation_systems/

[9] http://www.gpo.gov/fdsys/pkg/CFR-2011-title23-vol1/pdf/CFR-2011-title23-vol1-sec771-117.pdf.

[10] http://www.standards.its.dot.gov/Factsheets/Factsheet/71

[11] NHTSA 34-14 press release issued August 18, 2014; http://www.nhtsa.gov/About+NHTSA/Press+Releases/ NHTSA-issues-advanced-notice-of-proposed-rulemaking-onV2V-communications.

[12] http://ops.fhwa.dot.gov/publications/regitsarchguide/

[13] The Connected Vehicle Reference Implementation Architecture (CVRIA), a subset of the National ITS Architecture, defines functionality and information exchanges needed to provide connected vehicle applications, and supports connected vehicle project planning and development activities at State and local levels when an agency is ready to consider V2I deployment.

[14] Any entity seeking to install equipment or provide service related to V2I must comply with all applicable Federal Communications Commission (FCC) rules governing DSRC. See FCC Report and Order on DSRC, available at: (https://apps.fcc.gov/edocsjublic/attachmatch/FCC-03-324A1.pdf). Among other things, if an entity intends to deploy systems to provide V2I services under an FCC license, it must comply with all applicable FCC licensing rules and FCC rules implementing the National Environmental Policy Act and the National Historic Preservation Act.

[15] Connected Vehicle - Next Stage Certification Environment grant; http://www.federalgrants.com/ConnectedVehicle-Next-Stage-Certification-Environment-46654.html.

[16] http://ops.fhwa.dot.gov/publications/seitsguide/.

[17] FHWA Rule 940/FTA Policy, http://www.ops.fhwa.dot.gov/its_arch_imp/policy.htm.

[18] http://transition.fcc.gov/eb/mdrd/rules/pole.html.

[19] http://transition.fcc.gov/eb/mdrd/rules/pole.html.

[20] Ad hoc or non-critical maintenance activities would not be eligible for Federal-aid; however, this memorandum is collective and concise example of systematic "maintenance" programs, which are eligibility for Federal-aid under a preventive maintenance project. (http://www.fhwa.dot.gov/preservation/memos/011126.cfm)

[21] Connected Vehicle Reference Implementation Architecture (CVRIA) is a standards development plan that will allow identification and prioritization of the standards needed in support of connected vehicle implementation. http://www.standards.its.dot.gov/DevelopmentActivities/CVReference

[22] The purpose of Basic Safety Message standard, SAE J2735 DSRC Message Set Dictionary standard, is to support interoperability among safety, mobility, and sustainability applications through the use of standardized message sets, data frames and data elements. http://www.standards.its.dot.gov/Factsheets/Factsheet/71.

[23] 47 C.F.R. §15.5 is an existing FCC regulation to ensure that unlicensed devices do not cause interference with licensed users and require operators of unlicensed devices to immediately correct the problem or cease operation if interference occurs.

[24] http://wireless.fcc.gov/services/index.htm?job=service_home&id=dedicated_src

[25] http://wireless.fcc.gov/services/index.htm?job=licensing&id=dedicated_src.

[26] 47 C.F.R. §95.1501 - This subpart sets out the regulations governing Dedicated Short-Range Communications Service On-Board Units (DSRCS-OBUs) in the 5850-5925 MHz band. DSRCS Roadside Units (RSUs) are authorized under part 90 of this chapter and DSRCS, RSU, and OBU are defined in § 90.7 of this chapter.

[27] Refer to FHWA Operations Regulation and Policy http://www.ops.fhwa.dot.gov/opsregulationpolicy.htm.

In: Vehicle-to-Infrastructure Technologies ISBN: 978-1-63485-111-4
Editor: Albert Buchanan

Chapter 3

CRASH DATA ANALYSES FOR VEHICLE-TO-INFRASTRUCTURE COMMUNICATIONS FOR SAFETY APPLICATIONS*

Kimberly Eccles, Frank Gross, Mindy Liu and Forrest Council

FOREWORD

This report documents the results of crash data analyses to assess the potential safety benefits of vehicle-to-infrastructure (V2I) communication applications to improve highway safety. It provides estimates of the frequency and cost of crashes involving pre-crash scenarios addressed by V2I applications. It also evaluates pre-crash scenarios not addressed by those applications. This report will be useful to Federal, State, and local government agencies, research organizations, and private sector firms that research, develop, and deploy V2I technologies and safety applications.

Monique R. Evans
Director, Office of Safety Research and Development

---

* This is an edited, reformatted and augmented version of a contracted report, FHWA-HRT-11-040, issued by the U.S. Department of Transportation, Federal Highway Administration, November 2012.

ABBREVIATIONS

AIS	Abbreviated injury scale
AV	Autonomous vehicle
CICAS	Cooperative Intersection Collisions Avoidance System
CICAS-SLTA	Cooperative Intersection Collision Avoidance System—Signalized left-turn assist
CICAS-SSA	Cooperative Intersection Collision Avoidance System—Stop-sign assist
CICAS-V	Cooperative Intersection Collision Avoidance System—Violation
FARS	Fatality Analysis Reporting System
FHWA	Federal Highway Administration
GES	General Estimates System
HSIS	Highway Safety Information System
ITS	Intelligent Transportation System
KABCO	Scale used to record injury severity by crash victim
LTAP/OD	Left-turn across path/opposite direction
LVA	Lead vehicle accelerating
LVD	Lead vehicle decelerating
LVM	Lead vehicle moving at lower constant speed
LVS	Lead vehicle stopped
MAIS	Maximum abbreviated injury scale
NASS	National Automotive Sampling System
NHTSA	National Highway Traffic Safety Administration
PDO	Property damage only
SAS®	Statistical Analysis Software®
SCP	Straight crossing path
USDOT	United States Department of Transportation
UVC	*Uniform Vehicle Code*
V2I	Vehicle-to-infrastructure
V2V	Vehicle-to-vehicle
Volpe	John A. Volpe National Transportation Systems Center

SECTION 1. EXECUTIVE SUMMARY

The United States Department of Transportation (USDOT) Intelligent Transportation System (ITS) Strategic Plan has the potential to transform travel through safety, mobility, and environmental improvements in surface transportation. Two major programs under this plan are the Vehicle-to-Infrastructure (V2I) and Vehicle-to-Vehicle (V2V) Communications for Safety Initiatives. Both programs are composed of technologies and application areas that use wireless communication to enhance connectivity within the surface transportation network. This report focuses on the potential safety benefits of wireless communication between the roadway infrastructure and vehicles. Specifically, it identifies the magnitude, characteristics, and cost of crashes that would be targeted with currently proposed V2I for safety application areas that have been identified by USDOT with stakeholder input. It also identifies the magnitude, characteristics, and cost of the remaining crashes that are not targeted by a currently proposed V2I for safety application area (unaddressed crashes) for insight into potentially new applications or modifications to proposed applications. The V2I applications investigated in this study comprise four areas: (1) intersection safety, (2) speed management, (3) vulnerable road users, and (4) other safety application areas.

The primary analysis was conducted using the National Automotive Sampling System General Estimates System (NASS GES) database, which provides a national estimate of crashes by weighting a sample of crashes [1]. Other databases, including the Highway Safety Information System (HSIS) and the Fatality Analysis Reporting System (FARS), were used to supplement the NASS GES analysis [2, 3]. Pre-crash[†] scenarios, developed through previous efforts, were used to define each crash in the NASS GES database from 2005 to 2008 [1]. The frequency and severity of crashes in each pre-crash scenario were summarized, and total costs were assigned to each scenario based on the severity of crashes. Pre-crash scenarios were associated with the respective V2I application areas, and crash frequencies and associated costs were totaled within each application area to determine the potential safety benefits.

[†] There were 32 single-vehicle pre-crash scenarios and 44 multi-vehicle pre-crash scenarios for which there were observed crashes during the analysis period. Some of the pre-crash scenarios included the phrase "no maneuver," which refers to the driver's action before the crash, not if the driver maneuvered to avoid the crash. For example, if a motor vehicle turned right and then crashed into a pedestrian, it is considered a pedestrian/maneuver crash as opposed to a motor vehicle that was driving straight and then crashed into a pedestrian, which is considered a pedestrian/no maneuver crash.

Table 1 and table 2 present an overview of the findings from the NASS GES analysis for single-vehicle and multi-vehicle crashes. The tables provide annual data averaged over a 4-year period from 2005 to 2008. Specifically, they present the total annual crashes nationally, the subset of those crashes that would be targeted by current V2I application areas, and the remaining unaddressed crashes. Unaddressed crashes are calculated by subtracting the crashes that would be targeted by current application areas to those application areas from the total number of annual crashes. The associated cost for each group of crashes is also shown in the tables. Note that there is some overlap in the crashes targeted by the application areas. This overlap is accounted for in table 1 and table 2.

Table 1. Overview of estimates for targeted and unaddressed single-vehicle crashes identified in NASS GES [1]

Item	Estimated Annual National Crashes (Based on Weighted Data)	Estimated Cost
Total single-vehicle crashes	1,877,663	$164,132,235,633
Crashes potentially targeted by current application areas	1,106,966 (59 percent)	$120,078,331,482 (73 percent)
Unaddressed crashes	770,697 (41 percent)	$44,053,904,151 (27 percent)

Table 2. Overview of estimates for targeted and unaddressed multi-vehicle crashes identified in NASS GES [1]

Item	Estimated Annual National Crashes (Based on Weighted Data)	Estimated Cost
Total multi-vehicle crashes	4,099,936	$175,110,889,497
Crashes potentially targeted by current application areas	1,181,055 (29 percent)	$82,265,278,363 (47 percent)
Unaddressed crashes	2,918,881 (71 percent)	$92,845,611,134 (53 percent)

NASS GES estimates that approximately 6 million crashes (including both single- and multi-vehicle crashes) occurred each year from 2005 to 2008, totaling more than $339 billion in annual crash costs [1]. The total annual cost for single-vehicle crashes was $164 billion. The leading single-vehicle pre-crash scenarios were control loss/no vehicle action, road edge departure/no maneuver, and pedestrian/no maneuver. Collectively, these three scenarios represented 73 percent of the costs for single-vehicle crashes.

The total annual cost for multi-vehicle crashes was $175 billion. The leading multi-vehicle pre-crash scenarios were straight crossing path at non-signal, rear-end/lead vehicle stopped (LVS), and opposite direction/no maneuver. Collectively, these three scenarios represented 45 percent of the total costs for multi-vehicle crashes.

Currently identified V2I safety application areas could potentially target approximately

2.3 million crashes and $202 billion in costs. The remaining unaddressed crashes represent approximately 3.7 million crashes, totaling approximately $137 billion annually.

These unaddressed crashes represent potential targets for new V2I applications and are presented in this report by pre-crash scenario within single- and multi-vehicle categories. For single-vehicle crashes, pedestrian/no maneuver was the leading unaddressed pre-crash scenario, followed by bicyclist/no maneuver and animal/no maneuver. Collectively, these scenarios represented 56 percent of the costs of unaddressed single-vehicle crashes. From a practical perspective, V2I applications developed to mitigate these crashes would need to be carefully targeted due to the large number of potential miles for treatment.

The leading unaddressed multi-vehicle pre-crash scenario was rear-end/LVS, followed by straight crossing path at non-signal and left turn across path/opposite direction at non-signal. Collectively, these scenarios represented 50 percent of the costs of unaddressed multi-vehicle crashes. Although these scenarios represented a large potential target for V2I applications, a primary concern is whether these scenarios might be better addressed by autonomous vehicle (AV) or V2V communication applications.

The results indicated that the currently identified V2I safety applications discussed in this report are well conceived and can potentially treat large portions of the country's crashes and crash costs. However, there are many crashes that are not addressed by the applications analyzed in this report. The characteristics of unaddressed crashes that are presented in this report provide a starting point for identifying new applications or modifications to current applications.

SECTION 2. INTRODUCTION

2.1. Background

V2I communication for safety enables vehicles with 360-degree awareness to inform drivers of hazards and situations they cannot see. The following levels of action are envisioned:

- **Advisories**: A driver receives information in a non-time-critical manner.
- **Warnings**: An alarm signals to the driver that a crash is imminent and immediate action is required.

The V2I communication focuses on applications in which safety can be enhanced through connectivity that enables the exchange of information from a vehicle to the roadway infrastructure, from the infrastructure to a vehicle, or from the infrastructure to some other wireless-enabled device. Both original equipment and aftermarket solutions are being considered.

2.2. Scope

The objective of this study was to conduct and document crash data analyses of the primary V2I for safety application areas being evaluated by the Federal Highway Administration (FHWA) Safety Program. The goal was to determine which subsets of crashes are potentially treatable with currently identified V2I for safety application areas and which additional subsets could be treated either with modifications to the current application areas or with new application areas. The primary application areas of interest were intersection safety, speed, vulnerable road users, and others (i.e., applications that cannot be classified in the aforementioned areas of interest).

The specific objectives of the crash data analyses were as follows:

- Estimate the magnitude of the crash problem (i.e., number of crashes, fatalities, injuries, property damage, and the cost of the crashes) that could be impacted by currently identified V2I for safety application areas.
- Characterize the locations on the roadway for the crash types that could be impacted by currently identified V2I for safety application areas.
- Characterize pre-crash scenarios and contributing factors that would need to be addressed for currently identified V2I for safety application areas to be effective.
- Identify other significant crash types, pre-crash scenarios, and contributing factors, including currently identified and unidentified application areas, which might be amenable to solutions involving vehicle-infrastructure connectivity.
- Assess the relative potential safety benefits of currently identified and alternative V2I for safety application areas.

This report documents the most salient findings of this effort. It includes a description of the data used, the methods employed, and results by V2I application area. Supporting materials are provided in the appendix.

Throughout this report, crashes are described as either targeted crashes or unaddressed crashes with respect to V2I for safety application areas. Targeted crashes are crashes that could potentially be eliminated through the deployment of a specific V2I application or set of applications (i.e., researchers should determine the potential benefit of an application area, assuming 100 percent effectiveness and 100 percent deployment). The actual number of crashes mitigated depends on the effectiveness of the application and the extent of deployment. Unaddressed crashes are those that are not eliminated even if a V2I application, or set of applications, is 100 percent effective and fully deployed. That is not to say that unaddressed crashes cannot be mitigated by V2I application areas. Rather, currently identified application areas and potential extensions do not target these crashes. Unaddressed crashes might be covered through the development of new V2I application areas or by V2V and AV applications.

2.3. Methodology

The primary analysis involved an examination of the NASS GES database and several State databases from HSIS [1, 2]. These databases and analysis methodologies are briefly described in the following sections.

2.3.1. NASS GES Data Analysis

NASS GES contains data on a representative random sample of thousands of reported minor, serious, and fatal crashes involving passenger cars, pickup trucks, vans, large trucks, motorcycles, and pedestrians [1]. It is based on cases selected from a sample of police crash reports within randomly selected areas of the country.

The crash reports are chosen from 60 urban and rural areas that are representative of the geography, roadway mileage, population, and traffic density of the United States. Data collectors make weekly visits to approximately 400 police jurisdictions in the 60 areas throughout the United States and randomly sample approximately 50,000 police crash reports each year. Weights are provided so the sample data can be weighted to a national estimate. NASS GES data from 1988 to 2008 (crash, vehicle, and person files) are available online from the National Highway Traffic Safety Administration (NHTSA) [1].

This study analyzed the most recent 4 years of crash data from NASS GES, which was from 2005 through 2008. The analysis used the following five datasets for each year:

- **Accident:** This file contains data on crash characteristics and environmental conditions.
- **Vehicle:** This file contains data on the vehicles and drivers involved in the crash.
- **Person:** This file contains data on people involved in the crash, including age, gender, and injury severity.
- **Trafcon:** This file contains data on traffic control devices at the crash level.
- **Violatn:** This file contains data on violations charged to the driver(s) involved in the crash.

Analyses were conducted using both raw and imputed variables in the database. Raw data reflect the probability sample of police-reported crashes. The NHTSA report *Imputation in the NASS General Estimates System* describes imputation as the process of fabricating data when data are unknown [4]. Imputed variables are used to fill in unknown values. This is done because of historical precedence, convenience, consistency of data, and potential reduction in bias. Weighted variables are national estimates of crash characteristics based on weights established by NHTSA. According to the *NASS GES Analytical User's Manual 1988–2008*, the weight is the product of the inverse of the probabilities of selection at each of the three stages in the sampling process and is used to produce national estimates from the data [5]. Information on national estimates can be found in the *National Accident Sampling System General Estimates System Technical Note* [6].

2.3.1.1. Pre-Crash Scenarios

Crashes have multiple characteristics that can be grouped in an almost infinite number of ways. The research for this report used work conducted by The John A. Volpe National Transportation Systems Center (Volpe) to categorize each crash in the NASS GES database in a pre-crash scenario. The pre-crash scenarios were assigned using a Statistical Analysis Software® (SAS®) program developed by Volpe. Detailed criteria for assigning pre-crash scenarios are summarized in *Pre-Crash Scenario Typology for Crash Avoidance Research* [7]. Volpe classifies crashes by 38 single-vehicle pre-crash scenarios and 46 multi-vehicle pre-crash scenarios. During the 2005–2008 analysis period, there were observed crashes for 32 single-vehicle pre-crash scenarios and 44 multi-vehicle pre-crash scenarios. Distributions of crashes by pre-crash scenarios and vehicle type are presented in the appendix.

2.3.1.2. Vehicle Type

The Volpe SAS® program assigns one of six vehicle types to each vehicle involved in a crash. The vehicle types are light vehicle, transit vehicle, specialty vehicle, single-unit truck, combination-unit truck, and other. These vehicle types are assigned based on the NASS GES data variables for vehicle body type (bdytyp_h), special use type (spec_use), and trailer type (trailer) (see table 3). Definitions for vehicle body, special use, and trailer types can be found in the *National Automotive Sampling System (NASS) General Estimates System (GES) Analytical User's Manual 1988–2008* [5]. Motorcycles are included in the "other" and "specialty" vehicle types.

Table 3. Vehicle type assignment criteria

Vehicle Type	Assignment Criteria
Light vehicle	If (((1 ≤ bdytyp_h ≤ 22) or (28 ≤ bdytyp_h ≤ 41) or (45 ≤ bdytyp_h ≤ 49)) and spec_use = 0)
Transit vehicle	If (bdytyp_h in (25, 58, 59)) and (spec_use < 1 or spec_use = 3 or spec_use = 8 or spec_use = 9 or spec_use > 12)
Specialty vehicle	If ((80 ≤ bdytyp_h ≤ 89) or bdytyp_h in (23, 24, 42, 50, 65, 93, 97)) and ((4 ≤ spec_use ≤ 7) or spec_use = 2 or spec_use = 12)
Single-unit truck	If ((bdytyp_h in (60, 64, 66, 78, 79)) and (trailer in (1, 6)))
Combination-unit truck	If (bdytyp_h in (60, 64, 66, 78, 79)) and ((2 ≤ trailer ≤ 5))
Other	Else

Note: NASS GES data variables and codes used in these assignment criteria are defined in the National Automotive Sampling System (NASS) General Estimates System (GES) Analytical User's Manual 1988–2008 [5].

2.3.1.3. Crash Costs

Crash costs developed by Volpe were employed in this study as part of the economic analysis to ensure consistency between these results and those from past studies. All costs associated with NASS GES crash costs in this report were based on 2007 U.S. dollars.

The crash costs associated with each pre-crash scenario were calculated based on procedures used in a previous crash typology study, *Heavy Vehicle Pre-Crash Scenario Typology for Crash Avoidance Research* [8]. The conversion from crashes to economic costs was based on the severity of the crash. Specifically, DaSilva et al. used the maximum abbreviated injury scale (MAIS) [8]. The crash costs used in this study are based on MAIS (see table 4).

NASS GES does not provide detailed information regarding injury severity based on the MAIS coding scheme. Instead, it records injury severity by crash victim based on the KABCO scale as follows:

- **K:** Killed.
- **A:** Incapacitating injury.
- **B:** Non-incapacitating injury.
- **C:** Possible injury.
- **O:** No apparent injury.

Table 4. MAIS comprehensive crash costs (based on 2007 U.S. dollars) [8]

Consumer Price Index	Factor	MAIS 1	MAIS 2	MAIS 3	MAIS 4	MAIS 5	MAIS 6
1.346066	Medical	$3,204	$21,032	$62,585	$176,747	$447,509	$29,741
1.204077	Emergency medical services	$117	$255	$443	$999	$1,026	$1,003
1.277512	Market produce	$2,234	$31,960	$91,283	$135,977	$560,451	$760,577
1.277512	Household produce	$731	$9,354	$26,924	$35,782	$190,743	$244,696
1.204077	Insurance	$892	$8,319	$22,749	$38,934	$82,114	$44,695
1.277512	Workplace	$322	$2,495	$5,450	$6,002	$10,464	$11,117
1.204077	Legal	$181	$5,998	$19,034	$40,559	$96,153	$122,982
1.277512	Travel delay	$993	$1,081	$1,201	$1,276	$11,697	$11,687
1.204077	Property damage	$4,628	$4,761	$8,187	$11,840	$11,374	$12,369
1.277512	Quality-adjusted life years (QALYs)	$9,118	$186,525	$262,189	$784,777	$2,674,628	$4,889,799
New comprehensive costs		$22,420	$271,780	$500,045	$1,232,893	$4,086,149	$6,128,666
Injury subtotal		$16,799	$265,938	$490,657	$1,219,777	$4,063,088	$6,104,610
QALY relatives		0.0019	0.0381	0.0536	0.1605	0.547	1

Note: MAIS severity levels are as follows: MAIS 0 = no injury, MAIS 1 = minor, MAIS 2 = moderate, MAIS 3 = serious, MAIS 4 = severe, MAIS5 = critical, and MAIS 6 = fatal.

NASS GES also provides information for "injury severity unknown" and "died prior." Because there were not many "died prior" crashes, they were not considered in the analysis. DaSilva et al. used a conversion matrix to estimate MAIS injuries from the KABCO scale [8]. The series of multiplicative factors that were applied to convert injury severity from KABCO to MAIS designations were obtained from the NHTSA report, *Preliminary Regulatory Evaluation* [9].

The conversion factor was multiplied by MAIS subtotal dollar values in table 4 to obtain a weighted cost for each severity (travel delay and property damage are not included in this value).

For MAIS 0 (no injury), $2,423 was used based on unpublished values provided by Volpe. It should be noted that the goal of the current research was not to determine the exact dollar values associated with various crash types and pre-crash scenarios. Rather, the goal was to identify the relative magnitude of the crash problem for specific scenarios. As such, it was acceptable to use the unit costs in 2007 dollar values from the report by DaSilva et al. [8].

The resulting crash cost per maximum severity as reported in NASS GES was calculated using the procedure previously described and is shown in table 5. The crash cost per maximum severity was applied to the imputed crashes for each single-vehicle and multi-vehicle pre-crash scenario by maximum severity from 2005 to 2008.

Table 5. Cost per maximum severity (based on 2007 U.S. dollars)

NASS GES Code	Description	Cost
0	No injury	$4,597
1	Possible injury	$42,217
2	Non-incapacitating injury	$83,059
3	Incapacitating injury	$284,718
4	Fatal	$6,128,666
5	Injured, unknown severity	$118,770

For single-vehicle crashes, the total cost from 2005 to 2008 was $657 billion, with an annual average cost of $164 billion. The three highest-cost single-vehicle pre-crash scenarios represented 73 percent of the total costs for single-vehicle crashes ($478 billion of the $657 billion total) and included the following:

- Control loss/no vehicle action: $237 billion total and $59 billion annual average.
- Road edge departure/no maneuver: $168 billion total and $42 billion annual average.
- Pedestrian/no maneuver: $72 billion total and $18 billion annual average.

For multi-vehicle crashes, the total cost from 2005 to 2008 was $700 billion, with an annual average of $175 billion. The three highest-cost multi-vehicle pre-crash scenarios represented 45 percent of the total costs for multi-vehicle crashes ($315 billion of the $700 billion total) and included the following:

- Straight crossing path at non-signal: $131 billion total and $33 billion annual average.
- Rear-end/LVS: $96 billion total and $24 billion annual average.
- Opposite direction/no maneuver: $88 billion total and $22 billion annual average.

Detailed cost information for each pre-crash scenario by injury type is presented in the appendix (see table 24 for single-vehicle crashes and table 25 for multi-vehicle crashes).

2.3.1.4. Intersection Crashes

Many application areas required the identification of crashes related to various types of intersections and non-intersection segments. The location of the crash was characterized by the location where the first harmful event occurred. The *first harmful event* is defined as the occurrence of injury or damage involving a motor vehicle in transport, which can result from an impact or non-collision event. The variable imputed relation to junction (*RELJCT_I [A09I]*) specifies whether the crash occurred at a junction or non-junction area and whether it occurred at an interchange or non-interchange area. This variable was used to identify intersection-related and segment-related crashes.

Crashes were identified as either intersection, intersection-related, or segment crashes. Intersection crashes were crashes that occurred within the intersection, while intersection-related crashes occurred on the approach to or exit from an intersection and resulted from an activity, behavior, or control related to the movement of traffic through the intersection. Crashes were identified as intersection or intersection-related if the relationship to the junction was coded as any of the following:

- Non-interchange area, intersection (*RELJCT_I = 1*).
- Non-interchange area, intersection-related (*RELJCT_I = 2*).
- Non-interchange area, rail grade crossing (*RELJCT_I = 5*).
- Interchange area, intersection (*RELJCT_I = 11*).
- Interchange area, intersection-related (*RELJCT_I = 12*).

All other crashes were considered segment crashes, which included crashes that occurred at non-interchange areas (i.e., non-junctions, driveways or alley accesses, bridges, and crossovers) or at interchange areas (i.e., non-junctions, ramp exits/entrances, and other interchange locations). While crashes associated with driveways and other access points are similar to crashes at intersections, the V2I application areas discussed in this report did not target driveway crashes.

2.3.1.5. Location—Traffic Control and Alignment

Table 6 presents single-vehicle and multi-vehicle crashes by location. Intersection and intersection-related crashes were analyzed by type of traffic control, which was identified using the crash-level NASS GES data variable imputed traffic control device (*TRFCON_I[A16I]*) and, in some cases, the similarly named vehicle-level variable. Intersection crashes were classified by traffic control device as signalized intersection crashes (*TRFCON_I = 1*), stop-controlled intersection crashes (*TRFCON_I = 21*), and other intersection crashes. NASS GES determines the intersection control by the control affecting every vehicle in the crash. In a situation where two vehicles crash on an uncontrolled approach of a two-way, stop-controlled intersection, the intersection control for the crash is coded as uncontrolled.

Segment-related crashes were analyzed by presence of curvature (i.e., tangents [*ALIGN_I = 1*] versus curves [*ALIGN_I = 2*]). The majority of single-vehicle crashes occurred on tangent sections (67 percent) and curve sections (20 percent). The primary locations of multi-vehicle crashes were tangent sections (42 percent) and signalized intersections (28 percent).

Table 6. Distribution of crash location—average annual national crashes

Location of Crash		Single-Vehicle		Multi-Vehicle		Total	
		Crashes	Percent	Crashes	Percent	Crashes	Percent
Intersections and intersection-related	Signalized	67,520	4	1,147,720	28	1,215,240	20
	Stop-controlled	55,362	3	506,840	12	562,202	9
	Other intersections	122,171	7	518,938	13	641,109	11
Segments	Tangents	1,257,706	67	1,730,728	42	2,988,434	50
	Curves	374,904	20	195,710	5	570,614	10
Total		1,877,663	100	4,099,936	100	5,977,599	100

2.3.1.6. Area Type

Table 7 shows the assignment criteria and variable description for the area types defined in the NASS GES data. Area type was determined using the crash-level NASS GES data variable land use *(Land_Use[A05])*. Crashes that occurred within areas with a population greater than 25,000 people were considered urban, and areas reported as "other area" were considered rural.

Table 7. Area type assignment criteria

Area Type	Assignment Criteria	Variable Description
Urban	Land_use = 1	Area population of 25,000–50,000
	Land_use = 2	Area population of 50,000–100,000
	Land_use = 3	Area population of 100,000 or more
Rural	Land_use = 8	Other area
Unknown	Land_use = 9	Unknown

2.3.1.7. Speed-Related Crashes

Several application areas required the identification of crashes that were related to speeding. Speed-related crashes for single- and multi-vehicle pre-crash scenarios were identified using the NASS GES data speed-related variable (*SPEEDREL [D9N] = 1*). This variable was coded at the vehicle level to indicate whether speed was a contributing factor in the crash. If speed was coded as a factor for any of the involved vehicles, the crash was considered speed related.

2.3.1.8. Adverse Conditions

Some application areas required the identification of crashes that occurred during adverse driving conditions, which were defined based on the crash-level NASS GES data variables for roadway surface condition

(*Sur_Cond[A15]*) and weather (*Weather[A20]*). Crashes were considered to be related to adverse conditions if conditions for the crash were coded as any of the following:

- Wet (Sur_Cond = 2).
- Snow or slush (Sur_Cond = 3).
- Ice (Sur_Cond = 4).
- Sand, dirt, or oil (Sur_Cond = 5).
- Rain (Weather = 2).
- Sleet (Weather = 3).
- Snow (Weather = 4).
- Fog (Weather = 5).
- Rain and fog (Weather = 6).
- Sleet and fog (Weather = 7).

2.3.2. HSIS

HSIS is a roadway-based system maintained by FHWA that provides quality data on crash, roadway, and traffic variables linked to homogeneous sections of the highway system under State control [2]. It is the only multi-State database that allows for the safety analysis of roadway design factors through its file system and that has the capability to link roadway inventory and exposure data to crash data for a large sample of primary route mileage. It is also the only file system that includes both roadway sections with and without crashes. Currently, seven States are part of HSIS: California, Illinois, Maine, Minnesota, North Carolina, Ohio, and Washington. Historical data from Michigan and Utah are also available, but updated data are no longer captured. This study analyzed crash data for the most recent 3 years for California, Illinois, Minnesota, and Washington, which was from 2005 through 2007.

There are six types of data files available within HSIS, and all States maintain three basic files: a crash file, a roadway inventory file, and a traffic volume file. Additional roadway geometry files are also available within selected States, including a horizontal curve file (Illinois, Ohio, and Washington) and a vertical grade file (Illinois and Washington). Intersection and interchange data are also available for a limited number of States.

California and Minnesota were selected for detailed intersection analyses since these States provide intersection datasets. Illinois and Washington were used to conduct detailed analyses of curves and curve crashes since both States have a curvature file.

It is important to note that HSIS data are only available for State-maintained roadways in each State. As such, HSIS represents more rural than urban areas since roadways in urban areas are often maintained by a municipality. Quality data are largely unavailable for municipalities but would be helpful to better define the magnitude of the safety problems and potential impacts of V2I application areas in urban areas.

2.4. Previous Crash Typology Efforts Related to V2I for Safety

Previous studies have provided the foundation for this study, including the development of pre-crash scenarios and the investigation of the potential benefits of specific V2I for safety application areas. The following sections review select studies that were used as a basis for the analyses conducted as part of this current research. A brief overview of each study is provided, including the types of crash analyses performed and any gaps that were addressed as part of this project.

2.4.1. Pre-Crash Scenario Typology for Crash Avoidance Research

Najm et al. analyzed 2004 NASS GES data to develop a new typology of pre-crash scenarios for all police-reported crashes that involved at least one light vehicle (e.g., passenger car, sports utility vehicle, van, minivan, and light pickup truck) [7]. A total of 37 pre-crash scenarios were defined based on two existing typologies: the 44 crashes typology developed by General Motors® and the pre-crash scenarios typology developed by USDOT [10, 11]. Najm et al.'s new typology defines pre-crash scenarios that describe the vehicle movements, vehicle dynamics, and critical events that occurred immediately prior to the crash.

Each pre-crash scenario was ranked by three measures: crash frequency, functional years lost, and economic cost. Functional years lost and economic costs were estimated for each pre-crash scenario based on the severity of each crash assigned to that scenario. A summary table was provided for each pre-crash scenario that included the number of vehicles and people involved as well as the distribution of crashes by severity using the KABCO injury scale and the abbreviated injury scale (AIS). AIS uses a score of 1 (minor injury) through 6 (unsurvivable injury) to describe the injury severity of the crash victim. A detailed summary also described the typical scenario for a crash, factors that are overrepresented, dynamic variations of the scenario, and the general severity of crashes.

Single light-vehicle crashes resulted in an estimated economic cost of about $37 billion and 1.1 million functional years lost. The top three scenarios accounted for about two-thirds of all single light-vehicle crashes. In terms of economic costs and functional years lost, the top three pre-crash scenarios were as follows:

- Control loss without prior vehicle action (36.7 percent of economic costs and 38.4 percent of functional years lost).
- Road edge departure without prior vehicle maneuver (24 percent of economic costs and 24.7 percent of functional years lost).
- Pedestrian crash without prior vehicle maneuver (10.3 percent of economic costs and 12.6 percent of functional years lost).

Two-vehicle crashes involving at least one light vehicle resulted in an estimated economic cost of about $69 billion and 1.4 million functional years lost. The top three pre-crash scenarios accounted for about 40 percent of all two-vehicle crashes. In terms of economic costs, the top three scenarios were as follows:

- LVS: 14.9 percent.
- Vehicle(s) turning at non-signalized junctions: 10 percent.
- Straight crossing path at non-signalized junctions: 9.9 percent.

Multi-vehicle light-vehicle crashes (i.e., crashes involving more than two vehicles where at least one is a light vehicle) resulted in an estimated economic cost of about $14 billion and 292,000 functional years lost. The top three pre-crash scenarios accounted for 68 percent of all multi-vehicle crashes and were mostly related to rear-end crashes. In terms of economic costs, the top three scenarios were as follows:

- LVS: 35.9 percent.
- Lead vehicle decelerating (LVD): 14.8 percent.
- Opposite direction without prior vehicle maneuver: 7.1 percent.

These findings will help establish research priorities and provide a framework for a more consistent approach to identify interventions. The research also consolidates existing crash typologies into a single set of pre-crash scenarios from which all police-reported crashes can be categorized. While the study provided a basis for the current research, it was only based on a single year of data from 2004 and did not associate pre-crash scenarios or crash costs with specific V2I for safety application areas. The study also focused on light vehicles and did not specifically address crashes and costs related to other vehicle types (e.g., heavy vehicles and motorcycles). Crashes related to other vehicle types were included in the analysis but only when a single vehicle was involved. There is still a need to identify crashes and costs related to specific vehicle types.

2.4.2. Heavy Vehicle Pre-Crash Scenario Typology for Crash Avoidance Research

Similar to Najm et al.'s study on light vehicles, DaSilva et al. developed a new typology of pre-crash scenarios involving single-unit and combination-unit heavy vehicles (gross vehicle weight more than 10,000 lb) based on NASS GES data from 1996 through 2005 [8]. A total of 46 pre-crash scenarios were defined by describing the vehicle movements, vehicle dynamics, and critical events that occurred immediately prior to the crash.

Each pre-crash scenario was ranked by three measures: crash frequency, functional years lost, and economic cost. Functional years lost and economic costs were estimated for each pre-crash scenario based on the severity of each individual crash assigned to that scenario. A summary table was provided for each pre-crash scenario, including the number of vehicles and people involved and the distribution of crashes by severity using both KABCO and AIS. A detailed summary was also provided describing the typical scenario for each crash, factors that were overrepresented, dynamic variations of the scenario, and the general severity of the crashes.

Heavy vehicles accounted for approximately 6.5 percent (411,000 crashes) of all police-reported crashes annually. Of those, single-unit trucks accounted for an annual average of 214,000 crashes, and the remaining 197,000 crashes per year were associated with combination-unit trucks. The most common crash types for single-unit and combination-unit trucks were "offthe-roadway" and "changing lanes," respectively. Annually, approximately 974,000 people were involved in a crash with a heavy truck, and approximately 14 percent were injured or killed in those crashes. The annual economic cost of crashes involving heavy trucks, both single-unit and combination-unit, was estimated at $10.4 billion. More than 292,000 functional years were lost annually due to death and injury.

Single-unit truck crashes involving only one vehicle accounted for nearly 29 percent of all single-unit truck crashes, with an economic cost of approximately $877 million and 23,800 functional years lost. The top three pre-crash scenarios in terms of economic costs and functional years lost were as follows:

- Road edge departure/no maneuver.
- Pedestrian/no maneuver.
- Pedestrian/maneuver.

Two-vehicle single-unit truck crashes accounted for 67 percent of all single-unit truck crashes. These crashes resulted in an annual economic cost of $3.1 billion and 77,100 functional years lost. The top three pre-crash scenarios in terms of economic costs and functional years lost were as follows:

- Changing lanes/same direction.
- Turn at non-signal.
- Opposite direction/no maneuver.

Multi-vehicle (i.e., three or more vehicles) single-unit truck crashes accounted for 4 percent of all single-unit truck crashes. These crashes resulted in an annual economic cost of $458 million and 11,800 functional years lost. The top three pre-crash scenarios in terms of economic costs and functional years lost were as follows:

- Rear-end/LVS.
- Changing lanes/same direction.
- Opposite direction/no maneuver.

Single-vehicle combination-unit truck crashes represented 22 percent of all combination- unit truck crashes. These crashes resulted in an annual economic cost of $1 billion and 32,000 functional years lost. The top three pre-crash scenarios in terms of economic costs and functional years lost were as follows:

- Road edge departure/no maneuver.
- Control loss/no maneuver.
- Pedestrian/no maneuver.

Two-vehicle combination-unit truck crashes accounted for 73 percent of all combination- unit truck crashes. These crashes resulted in an annual economic cost of $4.3 billion and 125,400 functional years lost. The top three pre-crash scenarios in terms of economic costs and functional years lost were as follows:

- Changing lanes/same direction.
- Opposite direction/no maneuver.
- Turn at non-signal.

Multi-vehicle combination-unit truck crashes represented 5 percent of all combination-unit truck crashes. These crashes resulted in an annual economic cost of $712 million and 21,600 functional years lost. The top three pre-crash scenarios in terms of economic costs and functional years lost were as follows:

- Changing lanes/same direction.
- Rear-end/LVS.
- Rear-end/LVD.

These findings will help establish research priorities and provide a framework for a more consistent approach to identify countermeasures to target heavy truck crashes. The analysis identified the magnitude of the problem for 46 different pre-crash scenarios [5]. While the study provided a basis for the current research, it did not associate pre-crash scenarios or crash costs with specific V2I for safety application areas.

2.4.3. Frequency of Target Crashes for IntelliDrive Safety Systems

Najm et al. conducted a high-level analysis of potential collisions impacted by three general IntelliDrive safety systems [12]. The report employed the crash typologies developed in the *Pre-Crash Scenario Typology for Crash Avoidance Research* to estimate the potential safety effects of V2V, V2I, and AV communication systems [7]. The study looked at all light vehicles, heavy vehicles, and all vehicles combined. The analyses were based on NASS GES statistics from 2005 through 2008.

Target crashes were measured by the number of police-reported crashes that involved all vehicle types. Target crashes represented the maximum potential safety benefit if the fully deployed system was 100 percent effective in reducing target crashes. To avoid double counting, target crashes were first determined for a primary system category (e.g., V2V), and the remainder of the crash population was later assigned to the other two system categories (e.g., V2I and AV). Several analyses were reported, allowing each system to represent the primary countermeasure.

The study presented the results for each individual system and included tables and figures comparing the total percentage of crashes that could be targeted. Comparing each system individually as the primary countermeasure, V2V had the greatest potential to address crashes (4,409,000 target crashes), representing 74 percent of all police-reported crashes. AV systems represented the second greatest potential to address crashes (3,591,000 target crashes), accounting for 60 percent of all police-reported crashes. V2I was ranked third with respect to the potential to address total crashes (1,465,000 target crashes), representing 25 percent of all police-reported crashes.

The study also included results for the combination of the V2V and V2I systems. This combination accounted for the highest percentage of potential crashes targeted (4,503,000 crashes and 75 percent of all police-reported crashes).

The study also estimated potential benefits based on annual police-reported crashes involving at least one light vehicle and at least one heavy vehicle. Light vehicles include all passenger cars, vans, minivans, sports utility vehicles, and light pickup trucks with a gross vehicle weight rating less than 10,000 lb. Heavy vehicles included pickup, single-unit, and multi-unit trucks with a gross vehicle weight rating greater than 10,000 lb.

Results showed that the combination of V2V and V2I systems targeted the highest percentage of all light-vehicle crashes (77 percent) and crashes involving one heavy vehicle (71 percent). The V2V system alone targeted 76 percent of all light-vehicle crashes when it was considered the primary countermeasure. When considered individually as the primary countermeasure, AV systems targeted 60 percent of all light-vehicle crashes, and V2I systems targeted 25 percent of all light-vehicle crashes. The V2V system targeted 70 percent of all heavy-vehicle crashes when considered the primary countermeasure. When considered individually as the primary countermeasure, AV systems targeted 64 percent of all heavy-vehicle crashes, and V2I systems targeted 14 percent of all heavy-vehicle crashes.

These findings will help establish research priorities with respect to the various systems. While the study compared various systems in general, it did not associate pre-crash scenarios or crash costs with specific V2I for safety application areas.

2.4.4. Cooperative Intersection Collision Avoidance Systems Research on Comprehensive Costs of Intersection Crashes

Chang et al. estimated the magnitude of fatal and injury crashes as well as comprehensive crash costs associated with signalized and stop-controlled intersections [13]. Specifically, the study estimated the potential benefits of Cooperative Intersection Collision Avoidance Systems (CICAS) in terms of comprehensive crash costs. Three CICAS application areas were investigated: violation (CICAS-V), signalized left-turn assist (CICAS-SLTA), and stop sign assist (CICAS-SSA).

National estimates of crashes at signalized and stop-controlled intersections were identified based on data from FARS, NASS GES, and the Crashworthiness Data System [3, 1, 14]. Crashes were assigned to various crash types for both signalized and stop-controlled intersections corresponding to the CICAS application areas. For signalized intersections, crashes were assigned to violations (crossing path or non-crossing path) and signalized left-turn assist (left-turn across path/opposite direction (LTAP/OD) or left-turn and pedestrian). For stop-controlled intersections, crashes were assigned to violations (crossing path or non-crossing path) and stop sign assist (various crossing path crashes). Crash costs were assigned to each crash by allocating a unit cost to people and vehicles involved in a crash depending on four different categories: fatalities, injured people, non-injured people in injury vehicles, and property damage only (PDO). These costs were assigned based on the level of injury as indicated by MAIS.

In 2000, 43,000 fatalities were reported, with 9,500 occurring at intersections. In the same year, 2.7 million people were injured in vehicle crashes, of which, 1.3 million were related to crashes at intersections. A total of 8.9 million vehicles were involved in PDO crashes, of which, 4 million involved crashes at intersections. Intersection-related crashes and costs were further analyzed to identify the potential benefit of the CICAS application areas. The following results were reported for signalized intersections:

- CICAS-V (crossing path): $12 billion and 1,200 fatalities annually.
- CICAS-V (non-crossing path): $440 million and 100 fatalities annually.
- CICAS-SLTA (LTAP/OD): $9.1 billion and 420 fatalities annually.
- CICAS-SLTA (left-turn and pedestrian): $700 million and less than 100 fatalities annually.

The results for stop-controlled intersections were as follows:

- CICAS-V (crossing path): $6.2 billion and 1,300 fatalities annually.
- CICAS-V (non-crossing path): $600 million and 130 fatalities annually.
- CICAS-SSA (crossing path): $15 billion and 1,400 fatalities annually.

These findings will help establish research priorities with respect to the various CICAS systems. While the study compared specific CICAS systems, it did not identify the potential benefits of other V2I for safety application areas. The results also did not identify contributing factors or specific locations (e.g., urban/rural) where the CICAS applications could be most effective.

2.4.5. Infrastructure-Based Intersection Collision Avoidance Concept Study

A 2000 FHWA study explored the potential to reduce crossing path intersection crashes using advanced technology [15]. The primary objective of this study was to define and evaluate infrastructure-only concepts complementary to AV and V2I cooperative concepts to reduce the number of intersection crashes. High-priority intersection locations were chosen in three States (California, Minnesota, and Virginia), and a crash analysis was conducted at each location to identify the types of crossing path crashes and the potential causes of those crashes.

The three States provided the selection of high-priority intersections and hard copies of crash reports at these intersections. In total, 61 intersection locations were studied (21 locations in California, 20 locations in Minnesota, and 20 locations in Virginia). Crash data covered a 3-year period from 1997 through 1999 in California and from 1998 through 2000 in Minnesota and Virginia.

Crash reports showed that more than 50 percent of all crashes analyzed at intersection locations were crossing path crashes. Results from the crash analysis revealed that of all the crossing path crashes identified, LTAP/OD crashes were the predominant crash type at urban intersections, while straight cross path (SCP) crashes were the dominant crash type at rural intersections. As part of the crash analysis, two main causes of those crashes were identified: traffic control violations and insufficient gap. For insufficient gap crashes, the predominant causes were as follows:

- A total of 55 percent of drivers did not see other vehicle.
- A total of 15 percent of drivers thought the other vehicle would stop.

- A total of 11 percent of drivers had a view obstruction.
- A total of 9 percent of drivers misjudged the gap and thought they had a sufficient gap.
- A total of 5 percent of drivers did not look for oncoming vehicles.

The remaining 5 percent of the insufficient gap acceptance crashes were varied in their causes.

2.4.6. CICAS: Distribution of Crashes by Intersections—An Exploratory Analysis

A 2005 FHWA study investigated the magnitude and distribution of crossing path crashes at intersections [16]. A diverse set of data was used to conduct the analyses, including data from two States (Maine and California) and two cities (Detroit, MI, and San Francisco, CA). State-level crash and intersection data were obtained from HSIS. City-level data were obtained from a database developed as part of a previous effort; the data only included signalized intersections [2].

For the State-level analyses, the study identified the number of signalized and stop-controlled intersections by area type (i.e., rural or urban). The total number of crossing path crashes and crossing path crashes by severity for each intersection type and area type were also identified. In both States, crashes were overrepresented at signalized intersections. While the vast majority of intersections were stop-controlled (98 percent rural and 76 percent urban), the percentage of crashes at stop-controlled intersections was substantially less (61 percent rural and 56 percent urban). Crashes were also more prevalent in urban areas, accounting for 88 to 95 percent of crashes at signalized intersections and 57 to 65 percent of crashes at stop-controlled intersections.

For the city-level analyses, the study identified the number of crashes, injuries, and fatalities for various crossing path crash types, including left-turn, straight, and right-turn crossing paths. Left-turn crossing path was the leading crossing path crash type in both cities, followed by straight crossing path and right-turn crossing path. However, the severity was greatest for straight crossing path crashes.

While this study provided a basis for estimating the potential benefits of specific V2I for safety application areas, namely the CICAS application areas, it did not provide estimates at the national level. This study was also limited in that it only researched crossing path collisions and did not produce crash cost estimates for other scenarios.

2.4.7. Summary

Previous research efforts have defined pre-crash scenarios and quantified the relative safety issues in terms of crashes and economic costs. Research studies have explored potential safety impacts of V2I for safety applications. Some studies provided general comparisons of various ITS-related systems (i.e., V2I, V2V, and AV), while others focused on a specific subset of V2I applications (e.g., CICAS). While previous efforts have helped establish general research priority areas and have provided a foundation for subsequent efforts, they were either too general to identify benefits of specific applications or too specific (i.e., only focused on a subset of applications). Many of the studies focused on light vehicles with limited analysis of trucks, motorcycles, and pedestrians.

This study provides a comprehensive evaluation of the V2I for safety applications by the FHWA Safety Program to fill in gaps identified from the review of previous studies. Specifically, this report quantifies the national safety issue with respect to crashes and economic costs and shows the potential safety impacts of currently identified applications as well as the locations (e.g., rural/urban) where those applications could be deployed. It also identifies unaddressed crashes (i.e., those not targeted by current V2I applications). The results are presented in terms of all potential crashes targeted, with details for specific vehicle types and user groups when appropriate.

SECTION 3. OVERVIEW OF CURRENT V2I APPLICATION AREAS

The V2I application areas currently being considered by the FHWA Safety Program are characterized as intersection applications, speed applications, applications for vulnerable road users, and others.

Intersection applications are intended to prevent crashes at intersections and include the following:

- Drivers running red lights.
- Drivers running stop signs.
- Driver gap assist at signalized intersections.
- Driver gap assist at stop-controlled intersections.

Speed applications are intended to target crashes involving one or more vehicles when speeding contributed to the crash. These applications include the following:

- Curve speed warning.
- School zone speed warning.
- Work zone warning for reduced speed in work zones.
- Spot treatment/weather conditions.
- Speed zone warning.

Vulnerable road user applications are intended to target crashes involving users, such as pedestrians or vehicles, in vulnerable situations. These applications include the following:

- Work zone alerts.
- Infrastructure pedestrian detection.
- Priority assignment for emergency vehicle preemption.
- At-grade rail crossing.
- Bridge clearance warning.

Other applications that do not fit into the aforementioned categories include the following:

- Secondary accident warning.
- Lane departure warning.

Table 8 summarizes estimates that could be derived from a NASS GES-based analysis of the number of annual crashes that could be targeted by the current application areas. Note that it does not include data for several current application areas including school zone warning, emergency vehicle preemption priority, bridge clearance warning, and secondary accident warnings. The NASS GES system is based on a sample of police-reported crashes and is limited to the variables collected by the police. This restricts the ability to identify some of the circumstances of the crash that would be needed for these four applications. For example, the secondary accident warning application is intended to target crashes that occur, at least in part, following another crash (e.g., a rear-end crash in congestion that was caused by an earlier crash). This scenario is not specifically recorded by the police, and these crashes cannot be identified in NASS GES.

Table 8. Summary of annual estimated targeted crashes based on current applications

Application Area		Annual Single-Vehicle Crashes	Annual Multi-Vehicle Crashes
Intersection applications	Drivers running red lights	868	234,013
	Drivers running stop signs	3,586	40,838
	Driver gap assist at signalized intersections	N/A	200,212
	Driver gap assist at stop-controlled intersections	N/A	278,886
Speed applications	Curve speed warning	149,317	19,676
	Work zone warning for reduced speed	3,844	12,520
	Spot treatment/weather conditions	168,021	43,283
	Speed zone warning	154,339	206,356
Vulnerable road users applications	Work zone alerts	19,731	66,880
	Infrastructure pedestrian detection	17,812	N/A
	At-grade rail crossing	1,314	N/A
Other applications	Lane departure warning	1,041,460	195,187

N/A = Not applicable.

There is some overlap in the crashes targeted by the application areas, and some of these crashes can be targeted by more than one application. For example, some crashes can be targeted by the curve speed warning and the spot treatment/weather conditions applications.

The following sections provide information on the estimated annual crashes that may be targeted by each application area based on the analyses of NASS GES data. Crashes are described as targeted crashes or unaddressed crashes with respect to V2I application areas. Targeted crashes are crashes that could potentially be eliminated through the deployment of a specific V2I application or set of applications (i.e., after determining the potential benefit of an application area assuming 100 percent effectiveness and 100 percent deployment). The actual number of crashes addressed will depend on the effectiveness of the application and the extent to which they are deployed. Unaddressed crashes are those that would not be eliminated even if a V2I application or set of applications were 100 percent effective and fully deployed. Unaddressed crashes might be covered through the development of new applications or by other technologies.

SECTION 4. INTERSECTION APPLICATIONS

4.1. Running Red Light

This application area is intended to target crashes that result from signal violations. The running red light application provides a warning to drivers who are in danger of violating the signal.

4.1.1. Associated Pre-Crash Scenarios

The running red light application targets both the single-vehicle and multi-vehicle pre-crash scenario red light running. All of the crashes in this pre-crash scenario were considered potential targets for this application area.

4.1.2. Magnitude of Problem

There were an estimated 234,881 annual national target crashes based on weighted NASS GES data, and the estimated total annual cost of these crashes was more than $13 billion (note that annual refers to 2005 through 2008) [1]. Additionally, 45 percent of the target crashes resulted in fatalities or injuries.

4.1.3. Relevant Distributions

The distribution of target crashes by the six vehicle-type categories indicated that crashes involving two light vehicles represented 92 percent of the total crashes. All other vehicle types, including motorcycles, represented only a small portion (2 percent or less each) of the involved vehicles.

The distribution of target crashes by area type revealed that the majority of the crashes (78 percent) occurred in urban areas, and the remainder occurred in rural areas, indicating that signalized intersections are more prevalent in urban areas. The distribution of target crashes by posted speed limit indicated that the majority of the crashes (70 percent) occurred on roadways posted between 35 and 45 mi/h and were more prevalent in urban areas.

4.1.4. Relationships in HSIS Data

HSIS data from 2005 through 2007 were also investigated to determine the distribution of crash types by intersection traffic control [2]. While the NASS GES data indicated the magnitude of potential crashes impacted by the running red light application area, the HSIS analysis showed the magnitude of potential target crash types relative to other crash types at similar locations (i.e., the proportion of total signalized intersection collisions that could potentially be targeted by the application). There were 32,925 crashes identified at signalized intersections from the HSIS data in California from 2005 through 2007. The distribution by crash type showed that a large proportion of crashes at signalized intersections were broadside (angle), accounting for approximately 25.8 percent of crashes. A similar analysis was conducted for the 13,996 signalized intersection crashes from the HSIS data in Minnesota from 2005 through 2007. One of the leading crash types at signalized intersections was right-angle crashes, comprising 20.3 percent of signalized intersection crashes.

Of the 32,925 total crashes at signalized intersections in California from 2005 through 2007, 2,377 (7.2 percent) involved at least one heavy vehicle, and 775 (2.4 percent) involved at least one motorcycle. Fatal and severe injury crashes represented 1.9 percent of total crashes, 3.1 percent of heavy vehicle crashes, and 12.6 percent of motorcycle crashes at signalized intersections. Of those crashes involving at least one heavy vehicle at a signalized intersection, sideswipe crashes were the most prevalent (34.8 percent), followed by rear-end (28.6 percent) and broadside (18.9 percent). Of those crashes involving at least one motorcycle at a signalized intersection, broadside crashes were the most prevalent (28.4 percent), followed by rear-end (27.6 percent) and sideswipe (17.2 percent).

Of the 13,996 total crashes at signalized intersections in Minnesota from 2005 through 2007, 1,051 (7.5 percent) involved at least one heavy vehicle, and 153 (1.1 percent) involved at least one motorcycle. Similar to California, heavy vehicles and motorcycles were overrepresented in fatal and severe injury crashes at signalized intersections in Minnesota. Fatal and severe injury crashes represented 1.5 percent of total crashes, 2.3 percent of heavy vehicle crashes, and 11.1 percent of motorcycle crashes at signalized intersections. Of those crashes involving at least one heavy vehicle at a signalized intersection, rear-end crashes were the most prevalent (40.7 percent), followed by right-angle (18.9 percent) and sideswipe (18.8 percent). Of those crashes involving at least one motorcycle at a signalized intersection, rear-end crashes were the most prevalent (34.6 percent), followed by right-angle (18.3 percent) and other (18.3 percent).

4.2. Running Stop Sign

The running stop sign application area is intended to target crashes that result from stop sign violations at stop-controlled intersections. These intersections include two-way, four-way, and other stop-controlled intersections. The application provides a warning to drivers who are about to run a stop sign.

4.2.1. Associated Pre-Crash Scenarios

The running stop sign application area targets single-vehicle and multi-vehicle running stop sign pre-crash scenarios. All crashes in both scenarios are targeted for this application regardless of whether the intersection is two- or four-way stop-controlled. There are potentially other multi-vehicle crashes that result from a stop sign violation, but they were not coded as such by the police and are not included in these pre-crash scenarios. These crashes would have been included as straight crossing path at non-signals and turn at non-signals pre-crash scenarios. It is not possible to determine how many crashes in these pre-crash scenarios may be the result of stop sign violations versus other contributing factors, such as poor gap judgment; therefore, they were not counted in the estimate of targeted crashes. However, these pre-crash scenarios are targeted by the driver gap assistance at stop-control application addressed in section 4.4.

4.2.2. Magnitude of Problem

There were an estimated 44,424 annual national target crashes based on weighted NASS GES data, and the estimated total annual cost of these crashes was more than $2.0 billion. Additionally, 45 percent of all the target crashes resulted in fatalities or injuries.

4.2.3. Relevant Distributions

The distributions of target crashes by the six vehicle-type categories were considered for both single-vehicle and multi-vehicle crashes (multi-vehicle crashes referred to the first two vehicles involved). Crashes involving light vehicles represented the majority for both single-vehicle and multi-vehicle crashes. All other vehicle types represented only a small portion (2 percent or less each) of the involved vehicles.

The distribution of target crashes by area type indicated that the majority of multi-vehicle crashes (62 percent) occurred in urban areas, while the majority of single-vehicle crashes (59 percent) occurred in rural areas. This may highlight the need for different design considerations for applications in urban and rural areas.

4.2.4. Relationships in HSIS Data

HSIS data were investigated to determine the distribution of total crashes by intersection traffic control and area type [2]. While the NASS GES data indicated the magnitude of potential crashes impacted by this application, the HSIS analysis showed the prevalence of target intersections by area type and the magnitude of total crashes at those intersections. Based on the magnitude of total crashes at stop-controlled intersections, the HSIS intersection data analysis found that this application may be best targeted to two-way stop-controlled intersections. This was particularly true in rural areas, where crashes were more prevalent at two-way stop-controlled intersections compared to all-way stop-controlled intersections.

There were 35,758 crashes (51.7 percent of total) identified at two-way stop-controlled intersections from the HSIS data in California from 2005 through 2007. There were only 450 crashes (less than 1 percent of all intersection crashes) at all-way stop-controlled intersections in California. These results are based on total crashes (i.e., are not limited to specific crash types) and indicate the relative safety issue at two-way stop-controlled intersections when compared to all-way stop-controlled intersections.

A similar analysis identified 5,179 two-way stop-controlled intersection crashes (26.4 percent of total) from the HSIS data in Minnesota from 2005 through 2007. Similar to California, there were significantly fewer crashes (466 crashes, 2.4 percent of all intersection crashes) at all-way stop-controlled intersections. The distribution of two-way stop-controlled crashes in California and Minnesota by area type showed a large percentage (59.2 and 61.2 percent, respectively) of crashes in rural areas.

These results should be considered in the context of exposure. It may be expected that intersections with the greatest exposure (e.g., number of intersections and annual average daily traffic) will have the greatest number of crashes. In California and Minnesota, two-way stop-controlled intersections represented the greatest proportion of intersections from 2005 through 2007, accounting for 84 and 86 percent of all intersections in those States, respectively. All-way stop-controlled intersections accounted for less than 1 percent of all intersections by frequency and entering volume in both States. In California, both two-way and all-way stop-controlled intersections were more prevalent in rural areas, representing 73 percent of two-way stop-controlled intersections and 72 percent of all-way stop-controlled intersections. In Minnesota, two-way stop-controlled intersections were more prevalent in rural areas, accounting for 64 percent of all two-way stop-controlled intersections; however, all-way stop-controlled intersections were more prevalent in urban areas.

When compared to injuries at intersections with other traffic control types, two-way stop-controlled intersections in both California and Minnesota represented the greatest percentage of fatal and severe injury intersection crashes. In California, approximately 1.2 percent of two-way stop-controlled crashes resulted in fatalities, while 3.2 percent of these crashes resulted in severe injuries. In Minnesota, approximately 1.6 percent of two-way stop-controlled crashes resulted in fatalities, while 2.7 percent of these crashes resulted in incapacitating injuries.

Analysis of HSIS data by crash type showed a large number of angle and left-turn crashes at two-way stop-controlled intersections. The distribution by crash type in California indicated that 29.5 percent of crashes at two-way stop-controlled intersections were broadside (angle) type. California data did not include a specific crash type for left-turn crashes. Analysis of Minnesota data indicated that 49.7 percent of crashes at two-way stop-controlled intersections were right-angle crashes, and 6.8 percent were coded as left-turn crashes.

4.2.5. Potential Extension to Pedestrian and Bicycle Crashes

The running stop sign application area could potentially be extended to target vehicle-pedestrian crashes and vehicle-bicycle crashes at stop-controlled intersections. As a result, drivers would be warned of the presence of a crossing pedestrian or bicyclist, and pedestrians or bicyclists could be warned of a conflicting motor vehicle.

The single-vehicle pre-crash scenarios pedestrian/maneuver and bicyclist/maneuver were identified as target crashes for this application area. Specifically, the target crashes included crashes at stop-controlled intersections. Based on weighted GES data, there were an estimated 3,843 annual national target crashes, and the estimated total annual cost for these crashes was nearly $465 million.

4.3. Driver Gap Assist at Signalized Intersections

The driver gap assist at signalized intersections application area is intended to help drivers waiting to turn left at signalized intersections with permitted left turns through gap acceptance. The relevant crash type is a multi-vehicle crash involving a left-turning motor vehicle and a through motor vehicle.

4.3.1. Associated Pre-Crash Scenarios

This application targets the multi-vehicle pre-crash scenario left turn across path/opposite direction at signals. All of the crashes in this pre-crash scenario were considered potential targets for this application area. All crashes occurred at signalized intersections and involved a left-turning vehicle colliding with a through vehicle from the opposite direction. If the crash occurred because of a signal violation and not gap acceptance, it was included in a separate scenario.

4.3.2. Magnitude of Problem

There were an estimated 200,212 annual national target crashes based on weighted NASS GES data, and the estimated total annual cost of these crashes was more than $10.3 billion. Additionally, 44 percent of all the target crashes resulted in fatalities or injuries.

4.3.3. Relevant Distributions

Considering only the first two vehicles involved, the distribution of target crashes by the six vehicle-type categories indicated that crashes involving two light vehicles represented 94 percent of the total crashes. All other vehicle types accounted for only a small portion (2 percent or less each) of involved vehicles.

The distribution of target crashes by area type indicated that the majority of the crashes (77 percent) occurred in urban areas, and the remainder (23 percent) occurred in rural areas. This distribution reflects the prevalence of signalized intersections in urban areas.

4.3.4. Relationships in HSIS Data

Analysis of HSIS intersection data by area type confirmed the NASS GES analysis findings, which indicated the potential to impact a large number of signalized crashes in urban areas through the driver gap assist at signalized intersections application area despite the HSIS bias toward rural roads. There were 32,925 crashes identified at signalized intersections from the HSIS data in California from 2005 through 2007. The distribution by area type showed that crashes at signalized intersections occurred more frequently in urban areas: 83.5 percent of signalized intersection crashes occurred in urban areas compared to 16.5 percent in rural areas. A similar analysis was conducted for the 13,996 signalized intersection crashes from the HSIS data in Minnesota from 2005 through 2007. The majority of signalized intersection crashes were also in urban areas: 85 percent of signalized intersection crashes occurred in urban areas compared to 14.6 percent in rural areas. The remaining 0.4 percent of crashes occurred in area types coded as "other."

These results should be considered in the context of exposure. In California, signalized intersections were more prevalent in urban areas, representing 79 percent of intersections by number and 85 percent by entering volume. Additionally, 83.5 percent of crashes at signalized intersections occurred in urban areas. There was a similar trend in Minnesota, where signalized intersections in urban areas represented 83 percent of intersections by number and 79 percent by entering volume, as compared to 85 percent of the crashes.

4.4. Driver Gap Assist at Stop-Controlled Intersections

The driver gap assist at stop-controlled intersections application area is intended to target crashes that result from poor gap acceptance at two-way stop-controlled intersections. These crashes include stop-controlled motor vehicles that are traveling straight or turning at an intersection. *Gap acceptance*, in this case, is defined as the process by which a driver on the minor road stops at the stop sign and then makes a maneuver (right or left turn) onto or across the major road. This process requires the driver to judge the speed of conflicting traffic and the adequacy of "gaps" in traffic to complete the maneuver.

4.4.1. Associated Pre-Crash Scenarios

The driver gap assist at stop-controlled intersections application area targets straight crossing path at non-signal and turn at non-signal multi-vehicle pre-crash scenarios, which include crashes at unsignalized intersections and driveways. This application area is targeted to the subset of crashes that occur at two-way stop-controlled intersections. Multi-vehicle crossing path crashes at two-way stop-controlled intersections can be classified as resulting from gap acceptance or stop sign violations. This application is intended for crashes resulting from poor gap acceptance and not from stop sign violations. The assumption is that the majority of cases in which the investigating officer coded the crash as a violation would involve a vehicle that failed to stop at the sign rather than a vehicle that stopped and then proceeded. Crashes resulting from stop sign violations (e.g., a driver is ticketed for violating a stop sign) were categorized in a separate pre-crash scenario, running stop sign.

4.4.2. Magnitude of Problem

There were an estimated 278,886 annual national target crashes based on weighted NASS GES data, and the estimated total annual cost of these crashes was nearly $18.3 billion. Additionally, 38 percent of the target crashes resulted in fatalities or injuries.

4.4.3. Relevant Distributions

Considering only the first two vehicles involved, the distribution of target crashes by the six vehicle-type categories indicated that crashes involving two light vehicles represented 93 percent of the total crashes. All other vehicle types represented only a small portion (2 percent or less each) of the involved vehicles. Crashes involving motorcycles accounted for 2 percent of multi-vehicle target crashes, which is slightly greater than the occurrence of motorcycles in all multi-vehicle crashes. Motorcycles represented 1.4 percent of all multi-vehicle crashes.

The distribution of target crashes by area type indicated that the majority of the crashes (68 percent) occurred in urban areas. Roadways with five or more approach lanes represented 34 percent of the crashes. This is particularly noteworthy because it underscores the need for assistance with gap acceptance when crossing wider approaches. Roadways with two approach lanes represented 46 percent of the crashes, which can likely be attributed to the prevalence of this lane configuration.

4.4.4. Relationships in HSIS Data

HSIS data were investigated to determine the distribution of crash severity by intersection traffic control. This application area addresses potentially severe crashes at two-way stop-controlled intersections. The HSIS data support that these intersections represent a greater percentage of fatal and severe crashes when compared to other intersections. Of the 35,758 crashes at two-way stop-controlled intersections in California from 2005 through 2007, 1,543 crashes (4.4 percent) were fatal or severe injury. There were relatively fewer fatal and severe injury crashes at signalized and all-way stop-controlled intersections in terms of both frequency and percentage. Of the 32,925 crashes at signalized intersections, 639 crashes (1.9 percent) were fatal or severe injury. Of the 450 crashes at all-way stop-controlled intersections, 14 crashes (3.1 percent) were fatal or severe injury.

A comparable analysis was conducted using the HSIS data in Minnesota from 2005 through 2007. Similar to California, the frequency and percentage of fatal and severe injury crashes were greater at two-way stop-controlled intersections than at signalized and all-way stop-controlled intersections. Of the 5,179 crashes at two-way stop-controlled intersections, 222 crashes (4.3 percent) were fatal or severe injury. There were 13,996 crashes at signalized intersections and 466 crashes at all-way stop-controlled intersections during the same period. Of these, 218 crashes at signalized intersections (1.5 percent) and 5 crashes at all-way stop-controlled intersection (1.1 percent) were classified as fatal or severe injury crashes.

Analysis of HSIS intersection data by area type showed the potential for this application area to be better targeted to rural areas; nearly 60 percent of crashes at two-way stop-controlled intersections in California occurred in rural areas. Similarly, 61.2 percent of crashes at two-way stop-controlled intersections in Minnesota occurred in rural areas.

Heavy vehicles and motorcycles should be considered in the design of this application. More than 9 percent of the crashes at two-way stop-controlled intersections from the California HSIS data involved at least one heavy vehicle, more than a quarter of which were broadside crashes. Similarly, approximately 4 percent of the two-way stop-controlled crashes involved at least one motorcycle, more than 30 percent of which were broadside crashes. These heavy vehicle and motorcycle crashes were more likely to result in a fatality than other crashes at two-way stop-controlled intersections. Similar distributions were also identified in the Minnesota data for heavy vehicles and motorcycles.

SECTION 5. SPEED APPLICATIONS

5.1. Curve Speed Warning

The curve speed warning application area is intended to target crashes approaching horizontal curves on segments or interchange ramps that are speed-related. The application will provide a warning to drivers approaching a curve or ramp at an unsafe speed.

5.1.1. Associated Pre-Crash Scenarios

The curve speed warning application area targets several pre-crash scenarios including the following single-vehicle scenarios:

- Control loss/no vehicle action.
- Control loss/vehicle action when the vehicle action is not an intersection maneuver such as turning right or left.
- Road edge departure/no maneuver.
- Road edge departure/maneuver when the vehicle maneuver is not an intersection maneuver.
- Opposite direction/no maneuver.
- Rollover.
- Object contacted/no maneuver.

It also targets the following multi-vehicle pre-crash scenarios:

- Control loss/no vehicle action.
- Control loss/vehicle action when the vehicle action is not an intersection maneuver.
- Road edge departure/no maneuver.
- Opposite direction/no maneuver.

For both single-vehicle and multi-vehicle scenarios, speed-related crashes that occurred on a curve or an interchange ramp are targeted for this application.

5.1.2. Magnitude of Problem

There were an estimated 168,993 annual national target crashes based on weighted NASS GES data, and the estimated total annual cost of these crashes was more than $29 billion. This included 146,906 crashes at curve locations and 22,086 crashes at interchange ramps. Additionally, 44 percent of the target crashes resulted in fatalities or injuries.

5.1.3. Relevant Distributions

The distributions of target crashes by the six vehicle-type categories were considered for both single-vehicle and multi-vehicle crashes, with the latter referring to the first two vehicles involved. The overwhelming majority of the crashes involved light vehicles as expected. Vehicles in the "other vehicles" category represented 6 percent of the single-vehicle crashes, while trucks (single-unit and combination-unit) represented 12 percent of the multi-vehicle crashes. Additionally, 5 percent of the target single-vehicle crashes involved motorcycles, which was approximately twice the occurrence of motorcycles in all single-vehicle crashes (2.7 percent). Motorcycles were also overrepresented in multi-vehicle crashes, accounting for 4.3 percent of the target crashes but only 1.4 percent of all multi-vehicle crashes.

The distribution of target curve crashes by area type was considered for both single-vehicle and multi-vehicle crashes. The number of curve crashes was greater in urban areas for both crash types when compared to rural areas. This was somewhat unexpected because single-vehicle crashes are generally a greater concern in rural areas.

The distribution of crashes by number of lanes and land use indicated that two-lane roadways represented the majority of crashes for single-vehicle urban crashes (70 percent), single-vehicle rural crashes (85 percent), multi-vehicle urban crashes (63 percent), and multi-vehicle rural crashes (74 percent).

5.1.4. Relationships in HSIS Data

HSIS data were explored from 2005 through 2007 in Illinois and Washington to develop measures of exposure for curves by area type. While there were generally more curves and miles of curves in rural areas (possibly due to the rural sampling bias), the vehicle-miles traveled on curves were greater in urban areas. Traditionally, single-vehicle lane-departure crashes, often associated with curves, are a concern in rural areas. However, this concern is usually based on comparisons of this crash type to other crash types on rural roads (not in comparison to urban areas). While intersection crashes are traditionally thought of as the major urban safety concern, these results indicated that curve crashes in urban areas can also benefit from the curve speed warning application area.

HSIS data were also investigated to determine the distribution of curves by degree of curve and severity. While the GES data indicated the magnitude of potential crashes impacted by speed advisory and warning approaching horizontal curves, the HSIS data were used to help identify the curves where most of the crashes occurred.

Although the frequency of crashes is greater for tangents when compared to curves, an analysis of HSIS curve data revealed that crashes on curves tend to be more severe than those on tangent segments of the roadway. In Washington, there were 31,419 crashes identified on curves and 113,303 crashes identified on tangents from the HSIS data from 2005 through 2007. Approximately 0.9 percent of the curve crashes resulted in a fatality, 2.3 percent resulted in an incapacitating injury, and 10.9 percent resulted in a non-incapacitating injury. These percentages were higher than the respective percentages for crashes on tangents: 0.4 percent fatal crashes, 1.7 percent incapacitating injury crashes, and 8.9 percent non-incapacitating injury crashes by comparison.

Degree of curve measures the sharpness of the curve and is inversely related to curve radius (the higher the degree, the sharper the curve). The degree of curvature category with the most severe injury crashes was 10.01 to 20.0 degrees. For this category, 1.2 percent of crashes resulted in a fatality, 3.4 percent resulted in an incapacitating injury, and 13.9 percent resulted in a non-incapacitating injury.

5.2. Work Zone Warning for Reduced Speed in Work Zones

The work zone warning for reduced speed in work zones application area is intended to target speed-related crashes in work zones. Speeding drivers will be provided with a warning in active work zones.

5.2.1. Associated Pre-Crash Scenarios

Speed-related crashes are not specific to any particular pre-crash scenario but are found in multiple scenarios. All pre-crash scenarios were considered targets for this application. Work zones were the only speed zones that could be explicitly identified in the NASS GES data.

5.2.2. Magnitude of Problem

There were an estimated 16,364 annual national target crashes based on weighted NASS GES data, and the estimated total annual cost of these crashes was more than $1.3 billion. Additionally, 33 percent of the target crashes resulted in fatalities or injuries.

5.2.3. Relevant Distributions

Considering only the first two vehicles involved, the distribution of target crashes by the six vehicle-type categories indicated that crashes only involving light vehicles represented 86 percent of the total crashes. Additionally, 7 percent of single-vehicle target crashes involved combination-unit trucks, and 5 percent involved motorcycles. This was an overrepresentation of motorcycles by almost double because motorcycles only accounted for 2.7 percent of all single-vehicle crashes. All other vehicle types represented only a small portion (2 percent or less each) of the involved vehicles.

The distribution of target crashes by area type indicated that the majority of the crashes (55 percent) occurred in urban areas. Roadways with two lanes represented 46 percent of the crashes.

5.3. Spot Treatment/Weather Conditions

The spot treatment/weather conditions application area is intended to warn drivers to slow down for rain, ice, snow, or other adverse weather conditions that impact the roadway environment. These applications are envisioned to be applied to areas with a known history of weather-related crashes such as icy bridges or tangents in low-lying areas that flood during heavy rain.

5.3.1. Associated Pre-Crash Scenarios

The spot treatment/weather conditions application area targets crashes related to speeding during adverse weather conditions (based on both the surface conditions and the reported weather) for the single-vehicle and multi-vehicle pre-crash scenarios control loss/vehicle action and control loss/no vehicle action.

5.3.2. Magnitude of Problem

Based on weighted NASS GES data, there were an estimated 211,304 annual national target crashes, with an estimated total annual cost of $13.0 billion. Additionally, 28 percent of all the target crashes resulted in fatalities or injuries. (Note that these estimates include all locations, and targeting only selected locations decreases the estimates.)

5.3.3. Relevant Distributions

The distributions of target crashes by the six vehicle-type categories were considered for both single-vehicle and multi-vehicle crashes, with the latter referring to the first two vehicles involved. Vehicles in the "other vehicles" category represented 11 percent of the multi-vehicle crashes, while trucks (single- and combination-unit) represented 9 percent of the multi-vehicle crashes.

The distribution of target crashes by area type was considered for both single-vehicle and multi-vehicle crashes. The number of adverse weather crashes related to speeding was greater in urban areas for both single vehicle crashes (61 percent for urban) and multi-vehicle crashes (72 percent for urban) when compared to those in rural areas.

5.4. Speed Zone Warning

The speed zone warning application area is intended to warn drivers of reduced speed limits on tangents (e.g., on the approaches to small towns).

5.4.1. Associated Pre-Crash Scenarios

Speed-related crashes are not specific to any particular pre-crash scenario but are found in multiple scenarios. All pre-crash scenarios were considered targets for this application, and specific speed zones were used to identify target crashes. The NASS GES database does not explicitly identify speed zones. As a result, posted speed limits were analyzed for urban and rural locations based on statutory speed limits in the United States in section 11-802 of the *Uniform Vehicle Code* (UVC) [17]. UVC establishes a statutory speed limit of 55 mi/h in locations other than urban districts and 35 mi/h in urban districts. Roadways with posted speed limits lower than these statutory speed limits may have been established as "speed zones." In this study, posted speed limits of 30 mi/h and under and 50 mi/h and under were considered speed zones for urban and rural areas, respectively, excluding interstate highway locations. It is likely that using this method for identifying speed zones overestimates the number of speed zones, therefore overestimating the target crashes.

5.4.2. Magnitude of Problem

There were an estimated 360,694 annual national target crashes based on weighted NASS GES data, and the estimated total annual cost of these crashes was more than $28.5 billion. Additionally, 38 percent of the target crashes resulted in fatalities or injuries.

5.4.3. Relevant Distributions

Considering only the first two vehicles involved, the distribution of target crashes by the six vehicle-type categories indicated that crashes involving light vehicles represented 92 percent of single-vehicle crashes and 99 percent of multi-vehicle crashes. More than 4 percent of single-vehicle target crashes were motorcycles, which was an overrepresentation of motorcycles by almost double because they only accounted for 2.7 percent of all single-vehicle crashes. All other vehicle types represented only a small portion (2 percent or less each) of the involved vehicles.

The distribution of target crashes by area type indicated that the majority of the crashes (62 percent) occurred in rural areas. Roadways with two lanes represented more than half (56 percent) of the crashes.

SECTION 6. VULNERABLE ROAD USERS

6.1. Work Zone Alerts

The work zone alerts application area is intended to warn drivers of changes in traffic patterns due to construction (e.g., lane closures). It would be targeted to crashes in and approaching work zones. The application area is similar to the work zone warning for reduced speed in work zones but would not be limited to crashes involving speeding.

6.1.1. Associated Pre-Crash Scenarios

The work zone alerts application area targets crashes in and approaching work zones, including any crash occurring in or related to a work zone as identified by a work zone variable in the NASS GES database. The distribution of pre-crash scenarios indicates that several pre-crash scenarios are common in work zone related crashes. The pre-crash scenarios road edge departure/no maneuver (28 percent), control loss/no vehicle action (19 percent), and object contacted/no maneuver (14 percent) represented the largest proportions of single-vehicle work zone crashes. For multi-vehicle crashes, the leading pre-crash scenarios were rear-end/LVS (25 percent), changing lanes/same direction (16 percent), and rear-end/LVD (14 percent).

6.1.2. Magnitude of Problem

Based on weighted NASS GES data, there were an estimated 86,611 annual national target crashes, with an estimated total annual cost of more than \$4.5 billion. Additionally, 29 percent of all the target crashes resulted in fatalities or injuries.

6.1.3. Relevant Distributions

The NASS GES database provides information on the relation of the crash location to the work zone. According to the data, the majority of single-vehicle and multi-vehicle work zone crashes occurred in work or construction zones. Only 2 percent of the single-vehicle crashes and 5 percent of the multi-vehicle crashes did not occur in the work zone, but the first harmful event was related to the work zone. However, the location of a large proportion (43 percent) of work zone-related crashes (i.e., crashes in the work zone or on the approach) was not identified in the database.

The distribution of target crashes by the six vehicle-type categories was considered for both single-vehicle and multi-vehicle crashes, with the latter referring to the first two vehicles involved. While the majority of the crashes (86 percent) involved light vehicles, trucks (single- and combination-unit) represented 8 percent of both single-vehicle and multi-vehicle crashes. Motorcycles accounted for more than 3 percent of single-vehicle target crashes, which was higher than the percentage of motorcycle crashes for all single-vehicle crashes (1.4 percent), indicating that motorcycles were overrepresented in work zone crashes.

The distribution of target crashes by area type was considered for both single-vehicle and multi-vehicle crashes. The number of work zone crashes was greater in urban areas for both single-vehicle crashes (54 percent) and multi-vehicle crashes (63 percent) when compared to those in rural areas.

The distribution of target crashes by posted speed limit was considered for both single-vehicle and multi-vehicle crashes. It is unknown whether the data indicate the normal posted speed limits or the reduced work zone speed limits. A greater proportion of single-vehicle crashes occurred on higher speed limit roads (55 mi/h or greater) and low speed limit roads (25 mi/h or less) when compared to multi-vehicle crashes. At moderate speed limits (30–50 mi/h), multi-vehicle crashes represented a larger proportion of work zone crashes than single-vehicle crashes.

The distribution of target crashes by the number of lanes was considered for both single-vehicle and multi-vehicle crashes. The majority of work zone-related crashes occurred on roadways with three lanes or less (79 percent). Single-vehicle crashes represented a larger proportion of crashes on roadways with fewer lanes compared with multi-vehicle crashes.

6.2. Infrastructure Pedestrian Detection

The infrastructure pedestrian detection application area is intended to detect pedestrians and allows for changes in the walk phase of a signal (e.g., to allow more time for those with mobility impairments to cross). This application targets pedestrian crashes at signalized crossings including intersections and midblock crossings.

6.2.1. Associated Pre-Crash Scenarios

This application area targets single-vehicle and multi-vehicle pedestrian/maneuver and pedestrian/no maneuver pre-crash scenarios. Target crashes were limited to crashes that occurred at intersections and midblock locations with signalized control.

6.2.2. Magnitude of Problem

Based on weighted NASS GES data, there were an estimated 17,811 annual national target crashes, with an estimated total annual cost of more than $3.3 billion. Additionally, 99 percent of all the target crashes resulted in fatalities or injuries. Note that the estimated target is based on 506 observed crashes. As a result, it is less reliable than other estimates due to the small sample size.

6.2.3. Relevant Distributions

The relevant distribution of the single-vehicle target crashes is presented in this section. Only single-vehicle target crashes are included because they constitute the majority of the target crashes. The majority of single-vehicle target crashes occurred in urban areas (86 percent) compared to rural areas (14 percent).

According to the relationship to junction description in NASS GES, the target crashes mainly occurred at intersections (48 percent) or were intersection-related (51 percent) [1]. According to non-motorist location description in NASS GES, almost half of pedestrian crashes occurred in a crosswalk (49 percent) or in the roadway (43 percent) at the intersection [1]. Crashes at mid-block crossings, (i.e., not located at intersections) represented less than 7 percent of target crashes. It is likely that some of the crashes described as occurring in the intersection or intersection-related may have been signalized midblock crossings that were considered intersections. As a result, midblock crashes may be an underrepresented crash typology.

The distribution of crash types shows that more than half (53 percent) occurred between a pedestrian and a turning or merging vehicle. Light vehicles were involved in the majority (92 percent) of crashes. Only 2 percent of crashes involved transit vehicles, less than 2 percent involved trucks, and 4 percent of crashes involved other vehicles.

The distribution of lighting conditions for target crashes shows that although the majority of crashes occurred during daylight (65 percent), approximately 30 percent occurred during dark conditions (both dark (3 percent) and dark but lighted conditions (27 percent)).

Target crashes occurred most often on moderate speed limit roadways. A total of 20 percent were on roadways posted at 25 mi/h or less, more than 60 percent of crashes were roadways posted 30–35 mi/h, and 19 percent were roadways posted 40–45 mi/h.

Almost half of target crashes (48 percent) occurred on roads with six or more lanes, while less than one-fourth of the crashes were on one- or two-lane roads. This distribution reflects the increased potential for pedestrian crashes along wider, more vehicle-oriented roadways.

6.3. At-Grade Rail Crossing

This application area is intended to warn drivers of approaching trains, including light and heavy rail trains, for at-grade crossings.

6.3.1. Associated Pre-Crash Scenarios

The at-grade rail crossing application area targets single-vehicle and multi-vehicle crashes involving rail vehicles by providing warning of an approaching rail vehicle. This application is not limited by pre-crash scenario. Target crashes for this application include any crashes involving one or more vehicles and a train.

6.3.2. Magnitude of Problem

Based on weighted NASS GES data, there were an estimated 1,314 annual national target crashes, with an estimated total annual cost of more than $653 million. Additionally, 41 percent of all the target crashes resulted in fatalities or injuries. Note that the estimated target is based on 23 observed crashes. As a result, it is less reliable than other estimates. All but one of the observed crashes was a single-vehicle crash with a train.

6.3.3. Relevant Distributions

The distribution of traffic control type indicates that single-vehicle target crashes were split nearly evenly between active traffic control devices with gates, flashing lights, or traffic signals (50 percent) and passive devices including stop signs and crossbucks (46 percent). Few target crashes (less than 4 percent total) occurred at rail crossings without control or with other traffic control types.

The distribution of the six vehicle-type categories was considered for single-vehicle target crashes. While the majority of target crashes involved single vehicles (79 percent), 21 percent of crashes involved trucks, including single- and combination-unit trucks.

The distribution of the number of lanes for single-vehicle target crashes indicated that more than 80 percent of target crashes occurred on roadways with two or less lanes. Additionally, almost 90 percent of crashes occurred during favorable weather, while 8 percent of crashes occurred during rainy conditions. Finally, the distribution of land showed that the majority of single-vehicle crashes were in urban areas (80 percent), while only 20 percent occurred in rural areas.

SECTION 7. OTHER APPLICATION AREAS

Currently, there are two identified V2I for safety applications in the other application area category: secondary accident warning and lane departure warning. This section only describes the lane departure warning application. The secondary accident warning application could not be analyzed with NASS GES data because it is only a sample of crashes and not a comprehensive set in the database.

7.1. Lane Departure Crashes

The lane departure crashes application area is intended to warn drivers that they are about to unintentionally leave their lane or that they have unintentionally left their lane.

7.1.1. Associated Pre-Crash Scenarios

The application area targets single-vehicle and multi-vehicle pre-crash scenarios involving lane departure. It targets the same pre-crash scenarios as the speed curve warning application area; however, it is not limited to curve locations or speed-related crashes.

7.1.2. Magnitude of Problem

Based on weighted NASS GES data, there were an estimated 1,236,647 annual national target crashes, including more than 1 million single-vehicle crashes. The estimated cost of these crashes was more than $145 billion. Additionally, 35 percent of all the lane departure crashes resulted in fatalities or injuries.

7.1.3. Relevant Distributions

The majority of target crashes (92 percent) occurred on segments (62 percent occurred on tangent segments, and 30 percent occurred on curve segments). This is notable because lane departure crashes are not limited to curve locations.

While the majority of lane departure crashes occurred in urban areas (57 percent), more than 500,000 lane departure crashes occurred in rural areas, comprising the remaining 43 percent. Speed was a factor in more urban area crashes (18 percent) in comparison to rural areas (14 percent).

In total, 64 percent of all target crashes occurred on two-lane roads. Single vehicles had a slightly higher crash frequency (65 percent) than multi-vehicle crashes (59 percent).

SECTION 8. UNADDRESSED CRASHES

Crashes identified in the NASS GES data that are not targeted by one or more of the current V2I application areas mentioned previously are considered unaddressed crashes. These crashes can be explored to identify other potential applications. This section provides information on these crashes.

8.1. Distribution by Pre-Crash Scenarios

Table 9 and table 10 present the distribution and cost information of total and unaddressed crashes from NASS GES data sorted by pre-crash scenario for single-vehicle and multi-vehicle crashes, which are presented separately. The second row of table 9 indicates that for vehicle failure, unaddressed crashes represent 8 percent of total unaddressed crash costs, and 98 percent of the total crash cost for this scenario remains unaddressed. Pedestrian/no maneuver is the leading unaddressed single-vehicle pre-crash scenario, followed by cyclist/no maneuver and animal/no maneuver. Collectively, these three pre-crash scenarios represent 56 percent of the costs of unaddressed single-vehicle crashes. Unaddressed pedestrian, bicyclist, and animal pre-crash scenarios represent 67 percent of the unaddressed crash costs for single-vehicle crashes.

Table 9. Single-vehicle unaddressed crashes by pre-crash scenario

Pre-Crash Scenario	Total Crash Frequency	Unaddressed Crash Frequency	Unaddressed Cost (in millions of dollars)	Percent of Unaddressed Total Cost	Percent of Scenario Total Cost Unaddressed
1. No driver present	602	598	41	0	99
2. Vehicle failure	38,894	37,419	3,540	8	98
3. Control loss/vehicle action	77,173	24,145	1,446	3	23
4. Control loss/no vehicle action	457,714	0	0	0	0
5. Running red light	868	0	0	0	0
6. Running stop sign	3,586	0	0	0	0
7. Road edge departure/maneuver	85,141	56,215	1,718	4	51
8. Road edge departure/no maneuver	417,076	0	0	0	0
9. Road edge departure/backing	85,540	83,596	1,093	2	96
10. Animal/maneuver	17,844	17,735	172	0	99
11. Animal/no maneuver	295,063	291,390	3,936	9	95
12. Pedestrian/ maneuver	20,629	10,184	2,363	5	63
13. Pedestrian/ no maneuver	40,603	31,858	15,525	35	86
14. Cyclist/maneuver	21,022	20,629	2,222	5	99
15. Cyclist/ no maneuver	28,757	28,288	5,330	12	99
16. Backing into vehicle	371	371	4	0	100
18. Parking/same direction	163	147	10	0	83
19. Changing lanes/same direction	30,751	29,489	1,345	3	94

Table 9. (Continued)

Pre-Crash Scenario	Total Crash Frequency	Unaddressed Crash Frequency	Unaddressed Cost (in millions of dollars)	Percent of Unaddressed Total Cost	Percent of Scenario Total Cost Unaddressed
21. Opposite direction/maneuver	1,920	1,875	67	0	43
22. Opposite direction/no maneuver	25,593	0	0	0	0
23. Rear-end/striking maneuver	2,736	2,493	204	0	96
25. Rear-end/lead vehicle moving at lower constant speed (LVM)	2,131	1,746	201	0	92
26. Rear-end/LVD	7,771	6,828	674	2	92
27. Rear-end/LVS	8,057	7,498	284	1	90
31. SCP at non-signal	3,773	3,642	127	0	98
32. Turn at non-signal	7,572	7,193	315	1	94
33. Evasive maneuver/maneuver	3,624	3,486	148	0	97
34. Evasive maneuver/no maneuver	18,363	17,268	790	2	94
35. Rollover	6,126	0	0	0	0
36. Non-collision no impact	21,171	20,030	1,569	4	97
37. Object contacted/maneuver	68,220	66,574	930	2	92
38. Object contacted/no maneuver	78,808	0	0	0	0
Total	1,877,663	770,697	44,054	100	27

Note: Pre-crash scenarios for which there were no annual observed crashes were omitted from this table.

The distribution of unaddressed multi-vehicle crashes differed from the distribution of unaddressed single-vehicle crashes. Rear-end/LVS was the leading multi-vehicle pre-crash scenario followed by straight crossing path at non-signal and left turn across path/opposite direction at non-signal. Collectively, these three pre-crash scenarios represented 50 percent of the costs of unaddressed multi-vehicle crashes. The five rear-end pre-crash scenarios represented nearly 50 percent of the unaddressed multi-vehicle crashes and 44 percent of the unaddressed multi-vehicle crash costs. The Volpe report on frequency of target crashes suggests V2V and AV applications as the primary strategies for rear-end crashes [9].

The annual cost of unaddressed crashes was $136.9 billion, which includes all single-vehicle and multi-vehicle crashes. The leading unaddressed pre-crash scenarios are discussed in detail in the following sections.

Table 10. Multi-vehicle unaddressed crashes by pre-crash scenario

Pre-Crash Scenario	Total Crash Frequency	Unaddressed Crash Frequency	Unaddressed Cost (in millions of dollars)	Percent of Unaddressed Total Cost	Percent of Scenario Total Cost Unaddressed
1. No driver present	0	0	0	0	N/A
2. Vehicle failure	11,881	11,613	1,210	1	98
3. Control loss/vehicle action	31,332	534	8	0	1
4. Control loss/no vehicle action	63,080	0	0	0	0
5. Running red light	234,013	0	0	0	0
6. Running stop sign	40,838	0	0	0	0
7. Road edge departure/maneuver	1,479	1,394	162	0	97
8. Road edge departure/no maneuver	4,299	0	0	0	0
9. Road edge departure/backing	0	0	0	0	N/A
10. Animal/maneuver	404	404	8	0	100
11. Animal/no maneuver	6,065	5,645	344	0	93
12. Pedestrian/maneuver	646	641	22	0	98
13. Pedestrian/no maneuver	2,861	2,406	862	1	94
14. Cyclist/maneuver	155	89	7	0	73
15. Cyclist/no maneuver	500	438	96	0	99
16. Backing into vehicle	130,865	126,724	1,315	1	97
17. Turning/same direction	199,311	191,328	5,565	6	96
18. Parking/same direction	38,088	37,017	928	1	98
19. Changing lanes/same direction	309,933	295,994	6,189	7	96
20. Drifting/same lane	110,959	106,690	2,932	3	93
21. Opposite direction/maneuver	8,384	7,961	1,923	2	84
22. Opposite direction/ no maneuver	97,565	0	0	0	0
23. Rearend/striking maneuver	80,087	74,510	1,872	2	93
24. Rear-end/lead vehicle accelerating (LVA)	22,049	19,658	619	1	89
25. Rear-end/LVM	199,714	180,858	8,445	9	93
26. Rear-end/LVD	386,480	346,492	8,911	10	91

Table 10. (Continued).

Pre-Crash Scenario	Total Crash Frequency	Unaddressed Crash Frequency	Unaddressed Cost (in millions of dollars)	Percent of Unaddressed Total Cost	Percent of Scenario Total Cost Unaddressed
27. Rear-end/LVS	930,045	819,513	21,094	23	88
28. LTAP/OD at signal	200,212	0	0	0	0
29. Turn right at signal	30,980	30,251	957	1	97
30. Left turn across path/opposite direction at non-signal	182,574	178,663	11,260	12	95
31. SCP at non-signal	641,880	360,234	14,037	15	43
32. Turn at non-signal	38,788	29,229	570	1	81
33. Evasive maneuver/maneuver	8,936	8,317	269	0	93
34. Evasive maneuver/no maneuver	28,408	26,508	1,504	2	96
35. Rollover	450	443	409	0	100
36. Non-collision no impact	15,318	14,962	262	0	96
37. Object contacted/maneuver	595	507	29	0	92
38. Object contacted/no maneuver	9,220	8,755	369	0	96
39. Hit and run	3,551	3,547	64	0	91
40. Other—rear-end	1,516	1,427	19	0	98
41. Other— sideswipe	1,875	1,860	31	0	89
43 Other—turn across path	836	836	4	0	100
44 Other—turn into path	1,417	1,417	21	0	100
46. Other	22,347	22,016	527	1	97
Total	4,099,936	2,918,881	92,846	100	53

N/A = Not applicable.

Note: Pre-crash scenarios for which there were no annual observed crashes were omitted from this table.

8.2. Characteristics of Leading Unaddressed Crashes by Pre-Crash Scenario

8.2.1. Single-Vehicle Pedestrian Crashes

A total of 40 percent of unaddressed costs for single-vehicle crashes were represented by two pedestrian pre-crash scenarios, pedestrian/maneuver and pedestrian/no maneuver, with the latter being the more prevalent of the two. The estimated total annual cost of unaddressed single-vehicle pedestrian crashes was more than $17.8 billion. As stated in the Volpe report, these two crash scenarios are primarily targeted by AV applications, with the V2I applications only addressing crashes that occur at crosswalks [9].

The locations of these unaddressed pedestrian crashes are presented in table 11. There were no unaddressed pedestrian crashes at signalized intersections because those crashes are targeted by one of the currently identified applications and were previously discussed in this report. Intersections represented 35 percent of unaddressed single-vehicle pedestrian crashes, while the majority (58 percent) were non-junction crashes. The remaining crashes occurred at driveways/alleys (6 percent) and other locations. Additionally, 10 percent of the unaddressed pedestrian crashes occurred at stop-controlled intersections, which could potentially be targeted through the extension of the driver gap assist at stop-controlled intersections application as discussed in section 4.4 of this report.

Notably, the segment crashes represented the more severe crashes. Although segments represented 58 percent of the crashes, they represented 97 percent of the fatalities.

Table 11. Distribution of crash location for annual national (weighted) unaddressed edestrian crashes

Location of Crash		Crashes	Percent
Intersections	Signalized	0	0
	Stop-controlled	4,346	10
	Other Intersections	10,707	25
Segments	Tangents	23,331	55
	Curves	1,092	3
Driveways	Tangents	2,520	6
	Curves	47	0
Total		42,043	100

In total, 77 percent of these crashes occurred in urban areas. The majority of crashes also occurred on moderate speed roadways, with two-thirds of those crashes occurring on roadways with posted speeds limits of 25–35 mi/h. Crashes mainly occurred on two-lane roadways (41 percent).

8.2.2. Single-Vehicle Bicycle Crashes

In total, 17 percent of the total unaddressed costs for single-vehicle crashes were represented by two bicycle pre-crash scenarios, cyclist/maneuver and cyclist/no maneuver, with the latter being the more prevalent of the two.

The estimated total annual cost of unaddressed single-vehicle bicycle crashes was more than $7.5 billion. The primary strategies proposed for addressing cyclist crashes are AV applications, as reported in the Volpe study [9].

Table 12. Distribution of crash location for annual national (weighted) unaddressed bicycle crashes

Location of Crash		Crashes	Percent
Intersections	Signalized	12,520	26
	Stop-controlled	10,870	22
	Other Intersections	7,810	16
Segments	Tangents	9,429	19
	Curves	886	2
Driveways	Tangents	7,221	15
	Curves	181	0
Total		48,917	100

Table 12 presents the number and percentage of crashes based on location of unaddressed single-vehicle bicycle crashes. Intersection-related crashes represented 64 percent of unaddressed single-vehicle bicycle crashes, 21 percent of crashes were non-junction crashes, and 15 percent occurred at driveways/alleys.

Over 70 percent of the unaddressed bicycle crashes occurred in urban areas. The majority of crashes also occurred on moderate-speed roadways, with 70 percent of the crashes occurring on roadways with posted speeds limits of 25–35 mi/h. Additionally, 40 percent of the crashes occurred on two-lane roadways.

8.2.3. Single-Vehicle Animal Crashes

A total of 9 percent of the total unaddressed single-vehicle crashes were represented by two animal pre-crash scenarios, animal/maneuver and animal/no maneuver, with the latter being the more prevalent of the two. While 91 percent of the unaddressed animal crashes resulted in no injury, the estimated total annual cost of unaddressed animal crashes was more than $4.1 billion due to the large frequency of these crashes. Similar to cyclist crashes, AV applications are proposed as strategies for targeting animal crashes [9].

Over half of the unaddressed animal crashes (57 percent) occurred in rural areas, while the remaining 43 percent occurred in urban areas, indicating that animal crashes do not occur in just one area type (see table 13). The distribution of animal crashes by location (tangent and curve) and area type indicated that the crashes predominately occurred on tangent sections for both urban and rural locations (88 and 90 percent, respectively).

Table 13. Distribution of area type for annual national (weighted) unaddressed animal crashes

Area Type	Crashes	Percent
Urban	133,695	43
Rural	175,430	57
Total	309,125	100

The distribution of vehicle type for single-vehicle animal crashes indicated that light vehicles were involved in 95 percent of the crashes. Other vehicles accounted for 2 percent of the crashes. Each of the remaining vehicle types accounted for 1 percent or less of the total animal crashes.

8.2.4. Rear-End Crashes

A total of 44 percent of the total unaddressed multi-vehicle crashes were represented by five multi-vehicle rear-end pre-crash scenarios: rear-end/striking maneuver, rear-end/LVA, rear-end/ LVM, rear-end/LVD, and rear-end/LVS. The latter pre-crash scenario was the most prevalent of the five. The estimated total annual cost of unaddressed rear-end crashes was more than $40.9 billion. As previously mentioned, the majority of rear-end crashes are addressed by V2V and AV applications, which was a main cause for the high number of unaddressed read-end crashes in the Volpe report [9]. However, there were a substantial number of rear-end crashes targeted by the speed zone, work zone, and speed work zone V2I applications.

Table 14 presents the distribution of location for unaddressed crashes with urban and rural locations combined. Approximately half of the crashes occurred at intersection locations, and half occurred on roadway segments. The distribution by area type, shown in table 15, indicates that 76 percent of the crashes occurred in urban areas.

The distribution of vehicle type for multi-vehicle rear-end crashes indicated that crashes involving only light vehicles represented 93 percent of the crashes. Trucks, including single-unit and combination-unit vehicles, constituted 4 percent of rear-end crashes.

Table 14. Distribution of crash location for annual national (weighted) unaddressed rear-end crashes

Location of Crash		Crashes	Percent
Intersections	Signalized	442,037	31
	Stop-controlled	50,377	3
	Other intersections	199,728	14
Segments	Tangents	679,041	47
	Curves	69,848	5
Total		1,441,030	100

Table 15. Distribution of area type for annual national (weighted) unaddressed rear-end crashes

Area Type	Crashes	Percent
Urban	1,094,013	76
Rural	346,936	24
Unknown	81	0
Total	1,441,030	100

8.2.5. Straight Crossing Path Crashes at Non-Signals

A total of 15 percent of the total unaddressed multi-vehicle crashes were represented by the pre-crash scenario straight crossing path crashes at non-signals. The estimated total annual cost of these target crashes was more than $14 billion.

Table 16. Distribution of crash location for annual national (weighted) straight crossing ath crashes at non-signals

Location of Crash		Crashes	Percent
Intersections	Flashing signals	13,901	4
	Stop-controlled	61,258	17
	Other intersections	80,276	22
Segments	Tangents	10,162	3
	Curves	1,731	0
Driveways	Tangents	184,152	51
	Curves	8,754	2
Total		360,234	100

Table 16 presents the distribution of locations of unaddressed crashes. Notably, 53 percent of unaddressed straight crossing path at non-signal crashes occurred at driveways/alleys. The Volpe report proposes V2I applications for straight crossing path crashes at intersections and V2V applications for the straight crossing path crashes at non-intersections such as driveways [9]. Table 17 shows the distribution of these crashes by area type.

Table 17. Distribution of area type for annual national (weighted) straight crossing path crashes at non-signals

Area Type	Crashes	Percent
Urban	265,686	74
Rural	94,548	26
Total	360,234	100

The distribution of vehicle type for multi-vehicle, straight crossing path crashes at non-signals indicated that crashes only involving light vehicles represented 91 percent of crashes. Trucks, including single-unit and combination-unit vehicles, were involved in 4 percent of the crashes.

8.2.6. Left Turn Across Path/Opposite Direction Crashes at Non-Signals

The pre-crash scenario left turn across path/opposite direction at non-signals represented a total of 12 percent of the total unaddressed multi-vehicle crashes. The estimated total annual cost of these target crashes was more than $11.2 billion. The Volpe report proposes V2I applications only address intersection crashes for this pre-crash scenario, while V2V applications address all remaining crashes [9].

Table 18 presents the distribution of locations of unaddressed crashes. While the majority (60 percent) of the unaddressed left turn across path/opposite direction at non-signal crashes occurred at intersection or intersection related locations, 37 percent occurred at driveways/alleys.

A total of 18 percent of the crashes occurred on non-level roadways (i.e., on a grade, hillcrest, or sag). Over two-thirds (71 percent) of these unaddressed crashes occurred in urban areas, and two-thirds of the crashes occurred on roadways with a speed limit between 35 and 45 mi/h. The distribution of vehicle type for unaddressed multi-vehicle left turn across path/opposite direction crashes at non-signals indicated that crashes only involving light vehicles represented 96 percent of the crashes. Trucks, including single and combination-unit vehicles, were involved in less than 3 percent of the crashes.

Table 18. Distribution of crash location for annual national (weighted) unaddressed left turn across path/opposite direction crashes at non-signals

Location of Crash		Crashes	Percent
Intersections	Flashing signals	6,200	3
	Stop-controlled	16,554	9
	Other intersections	86,236	48
Segments	Tangents	3,932	2
	Curves	154	0
Driveways	Tangents	62,171	35
	Curves	3,415	2
Total		178,663	100

8.3. Characteristics of Unaddressed Crashes by Vehicle, Area, and Location Type

8.3.1. Vehicle Type

The distribution of unaddressed single-vehicle and multi-vehicle crashes by vehicle type is presented in table 19. For single-vehicle crashes, single-unit and combination-unit trucks appear to be overrepresented, accounting for 8 percent of the vehicles. Other vehicles also are overrepresented when compared to multi-vehicle crashes, representing another 6 percent of the unaddressed crashes.

Table 19. Distribution of vehicle type for unaddressed crashes

	Single-Vehicle Crashes		Multi-Vehicle Crashes		Total	
Vehicle Type	Number of Vehicles Involved	Percent	Number of Vehicles Involved	Percent	Number of Vehicles Involved	Percent
Light vehicles	666,118	86	5,430,129	93	6,096,247	92
Transit vehicles	2,334	0	24,471	0	26,805	0
Special vehicles	3,466	0	15,965	0	19,431	0
Single-unit trucks	35,057	5	112,382	2	147,439	2
Combination-unit trucks	20,101	3	106,979	2	127,080	2
Other vehicles	44,457	6	100,619	2	145,076	2
Not reported	0	0	47,216	1	47,216	1
Total	771,533	100	5,837,761	100	6,609,294	100

8.3.2. Area Type

Table 20 presents the distribution of unaddressed crashes by area type for single-vehicle and multi-vehicle crashes. More than 70 percent of the unaddressed crashes were in urban areas. Urban locations represented a greater proportion of the unaddressed crashes for multi-vehicle crashes than for single-vehicle crashes.

Table 20. Distribution of area type for unaddressed crashes

Area	Single-Vehicle		Multi-Vehicle		Total	
Type	Crashes	Percent	Crashes	Percent	Crashes	Percent
Urban	454,274	59	2,171,556	74	2,625,829	71
Rural	317,164	41	747,152	26	1,064,316	29
Unknown	95	0	173	0	268	0
Total	771,533	100	2,918,881	100	3,690,413	100

8.3.3. Location Type

Table 21 presents the distribution of unaddressed crashes by intersection and segment locations. For both single-vehicle and multi-vehicle crashes, the majority of the unaddressed crashes were on tangent segments. This was expected because many of the current applications target intersections and curved segments. Specifically, 75 percent of single-vehicle crashes compared to 51 percent of multi-vehicle crashes occurred on tangents. Furthermore, the difference in percentage between single-vehicle and multi-vehicle crashes for crashes at intersections and segments was expected because multi-vehicle crashes are generally more prevalent at intersections. Additionally, 21 percent of the unaddressed multi-vehicle crashes were at signalized intersections. Signalized intersections could be a target for new applications or extensions to existing application areas.

Table 21. Distribution of crash location for unaddressed crashes

Location of Crash		Single-Vehicle		Multi-Vehicle		Total	
		Crashes	Percent	Crashes	Percent	Crashes	Percent
Intersections	Signalized	31,800	4	613,532	21	645,332	17
	Stop-controlled	27,835	4	163,404	6	191,239	5
	Other intersections	73,829	10	512,332	18	586,161	16
Segments	Tangents	575,672	75	1,500,622	51	2,076,294	56
	Curves	62,397	8	128,990	4	191,387	5
Total		771,533	100	2,918,881	100	3,690,413	100

SECTION 9. CONCLUSION

This report documents the annual national frequencies and costs of crashes that could be targeted for 12 of the currently identified V2I for safety application areas. These analyses also provide information on crash scenarios and the associated costs that are unaddressed by these 12 safety applications. These unaddressed crashes can be explored to identify other potential applications.

Table 22 summarizes the estimated annual crashes and associated costs that could be targeted by the 12 V2I for safety application areas. These application areas could potentially target a significant number of crashes and their associated crash costs (more than $200 billion annually). When combined, the costs would total more than $200 billion, and the overlap among some of the application areas is accounted for in the total. The lane departure application represents the greatest potential to improve safety with respect to crash frequency and the potential crash cost. This finding is expected because the targets for this application are several pre-crash scenarios with significant crashes resulting from vehicles leaving their lanes.

Table 22. Overview of annual target crashes and associated costs identified in NASS GES for currently identified application areas

Application Area		Estimated Annual Crashes Targeted	Annual Cost of Crashes Targeted (millions of dollars)
Intersection applications	Running red light	234,881	13,152
	Running stop sign	44,424	2,034
	Driver gap assist at signalized intersections	200,212	10,252
	Driver gap assist at stop-controlled intersections	278,886	18,273
Speed applications	Curve speed warning	168,993	29,080
	Work zone warning for reduced speed	16,364	1,335
	Spot treatment/weather conditions	211,304	13,019
	Speed zone warning	360,695	28,500
Vulnerable road users applications	Work zone alerts	86,611	4,563
	Infrastructure pedestrian detection	17,812	3,333
	At-grade rail crossing	1,314	653
Other applications	Lane departure warning	1,236,647	145,347
Total (accounting for overlaps)		2,288,021	202,344

Some caution is necessary because these results represent all potential target crashes for each application area. Targeted crashes are those that could potentially be eliminated through the deployment of a specific V2I application or set of applications (assuming 100 percent effectiveness and 100 percent deployment). The actual number of crashes addressed depends on the effectiveness of the application and the extent to which it is deployed.

The 12 analyzed application areas have the potential to cover approximately 73 percent of the single-vehicle total crash costs and 47 percent of the multi-vehicle total crash costs (see table 1 and table 2). The remaining crashes are considered unaddressed crashes, which are crashes that would not be eliminated even if the analyzed V2I applications were 100 percent effective and fully deployed. Unaddressed crashes might be covered through the development of new application areas or by other V2I technologies. They account for $44.0 billion in single-vehicle crash costs and $92.8 billion in multi-vehicle crash costs.

The following currently identified V2I for safety application areas could not be explored due to small samples or limitations of the NASS GES data:

- School zone speed warning.
- Priority assignment for emergency vehicle preemption.
- Bridge clearance warning.
- Secondary accident warning.

Other datasets and methods would be needed to identify the target populations for these application areas and estimate costs associated with these crash types.

Based on the percentage of crash costs targeted, current application areas appear to be well conceived in that they target major crash scenarios, particularly for single-vehicle crashes. In addition, a relatively large percentage of unaddressed crash costs are related to rear-end collisions. While V2I application areas could be conceived to target these scenarios, the Volpe report suggests V2V and AV as the primary applications areas for addressing rear-end crashes [12].

More detailed crash data analysis is necessary to help target the deployment of V2I applications to identify the maximum benefit per dollar spent. For example, the benefits of the pre-crash scenario driver gap assist at signalized intersections would be expected to be greater at intersections with permissive left-turn phases and high numbers of potential conflicts (e.g., high volume of left turn and opposing through vehicles).

Addendix A: Relevant Distributions

Table 23. Frequency of annual observed and annual weighted single-vehicle and multi-vehicle crashes by pre-crash scenario

Pre-crash Scenario	Single-Vehicle Crashes		Multi-Vehicle Crashes	
	Annual Observed Crashes	Annual Weighted Crashes	Annual Observed Crashes	Annual Weighted Crashes
1. No driver present	7	602	0	0
2. Vehicle failure	467	38,894	184	11,881
3. Control loss/vehicle action	893	77,173	350	31,332
4. Control loss/no vehicle action	5,620	457,714	984	63,080
5. Running red light	18	868	2,932	234,013
6. Running stop sign	41	3,586	478	40,838
7. Road edge departure/maneuver	656	85,141	30	1,479
8. Road edge departure/no maneuver	4,693	417,076	90	4,299
9. Road edge departure/backing	404	85,540	0	0
10. Animal/maneuver	101	17,844	3	404
11. Animal/no maneuver	1,686	295,063	52	6,065
12. Pedestrian/maneuver	562	20,629	7	646
13. Pedestrian/no maneuver	1,226	40,603	40	2,861
14. Cyclist/maneuver	564	21,022	3	155
15. Cyclist/no maneuver	716	28,757	9	500
16. Backing into vehicle	3	371	679	130,865
17. Turning/same direction	0	0	1,562	199,311
18. Parking/same direction	4	163	256	38,088
19. Changing lanes/same direction	392	30,751	2,750	309,933
20. Drifting/same lane	0	0	1,055	110,959
21. Opposite direction/maneuver	18	1,920	118	8,384

Pre-crash Scenario	Single-Vehicle Crashes		Multi-Vehicle Crashes	
	Annual Observed Crashes	Annual Weighted Crashes	Annual Observed Crashes	Annual Weighted Crashes
22. Opposite direction/no maneuver	248	25,593	1,228	97,565
23. Rear-end/striking maneuver	26	2,736	577	80,087
24. Rear-end/LVA	0	0	151	22,049
25. Rear-end/LVM	34	2,131	1,861	199,714
26. Rear-end/LVD	119	7,771	3,078	386,480
27. Rear-end/LVS	97	8,057	6,853	930,045
28. LTAP/OD at signal	0	0	2,359	200,212
29. Turn right at signal	0	0	228	30,980
30. Left turn across path/opposite direction at non-signal	0	0	1,903	182,574
31. SCP at non-signal	39	3,773	5,783	641,880
32. Turn at non-signal	97	7,572	242	38,788
33. Evasive maneuver/maneuver	38	3,624	71	8,936
34. Evasive maneuver/no maneuver	217	18,363	291	28,408
35. Rollover	149	6,126	13	450
36. Non-collision no impact	191	21,171	133	15,318
37. Object contacted/maneuver	326	68,220	9	595
38. Object contacted/no maneuver	615	78,808	100	9,220
39. Hit and run	0	0	26	3,551
40. Other—rear-end	0	0	13	1,516
41. Other—sideswipe	0	0	13	1,875
43. Other—turn across path	0	0	5	836
44. Other—turn into path	0	0	9	1,417
46. Other	0	0	145	22,347
Total	20,265	1,877,663	36,668	4,099,936

Note: Pre-crash scenarios for which there were no annual observed crashes were omitted from this table.

Table 24. Cost of annual single-vehicle crashes by pre-crash scenario and injury type

Pre-Crash Scenario	Cost by Injury Type (millions of dollars)						
	No Injury	Possible Injury	Non-Incapacitating	Incapacitating	Fatal	Injured, Unknown Severity	Total
1. No driver present	1	3	11	15	0	11	41
2. Vehicle failure	119	199	447	690	2,143	24	3,622
3. Control loss/vehicle action	247	358	847	1,102	3,730	39	6,323
4. Control loss/no vehicle action	1,299	2,667	5,511	10,277	39,240	362	59,356
5. Running red light	2	5	22	22	0	0	51
6. Running stop sign	11	16	61	49	18	0	155
7. Road edge departure/maneuver	323	281	469	571	1,677	37	3,358
8. Road edge departure/no maneuver	1,261	2,177	4,603	7,834	25,768	478	42,121
9. Road edge departure/backing	375	76	109	141	396	36	1,135
10. Animal/maneuver	76	42	24	29	0	4	174
11. Animal/no maneuver	1,230	528	896	1,116	333	28	4,131
12. Pedestrian/maneuver	3	354	586	973	1,737	87	3,740
13. Pedestrian/no maneuver	6	515	1,155	2,797	13,348	149	17,969
14. Cyclist/maneuver	9	283	782	571	507	97	2,249
15. Cyclist/no maneuver	13	419	923	1,183	2,808	34	5,379
16. Backing into vehicle	2	0.04	2	0	0	0	4
18. Parking/same direction	0.25	2	3	7	0	0	12
19. Changing lanes/same direction	91	222	343	420	340	14	1,429
21. Opposite direction/maneuver	5	17	24	18	91	0	155
22. Opposite direction/ no maneuver	80	153	224	476	498	3	1,434
23. Rear-end/striking maneuver	10	11	25	17	149	0	212
25. Rear-end/LVM	6	10	37	57	108	0	218

Pre-Crash Scenario	Cost by Injury Type (millions of dollars)						
	No Injury	Possible Injury	Non-Incapacitating	Incapacitating	Fatal	Injured, Unknown Severity	Total
26. Rear-end/LVD	22	37	118	150	400	7	734
27. Rear-end/LVS	25	51	85	121	25	9	315
31. SCP at non-signal	11	30	41	48	0	0	129
32. Turn at non-signal	23	32	107	173	0	2	336
33. Evasive maneuver/maneuver	12	16	31	56	35	1	152
34. Evasive maneuver/no maneuver	56	112	202	300	163	10	843
35. Rollover	8	51	148	324	921	3	1,455
36. Non-collision no impact	77	55	165	275	1,031	9	1,611
37. Object contacted/maneuver	297	82	86	91	437	19	1,012
38. Object contacted/ no maneuver	315	173	288	554	2,913	36	4,279
Total	6,013	8,973	18,375	30,455	98,817	1,499	164,132

Note: Pre-crash scenarios for which there were no annual observed crashes were omitted from this table.

Table 25. Cost of annual multi-vehicle crashes by pre-crash scenario and injury type

	Cost by Injury Type (millions of dollars)						
Pre-Crash Scenario	No Injury	Possible Injury	Non-Incapacitating	Incapacitating	Fatal	Injured, Unknown Severity	Total
2. Vehicle failure	37	74	90	215	810	9	1,235
3. Control loss/vehicle action	104	212	213	249	660	10	1,448
4. Control loss/no vehicle action	172	533	648	1,133	4,755	45	7,286
5. Running red light	593	2,556	2,473	3,373	3,835	270	13,099
6. Running stop sign	102	414	519	543	223	80	1,882
7. Road edge departure/maneuver	4	11	14	17	118	3	168
8. Road edge departure/no maneuver	11	25	59	125	604	3	828
10. Animal/maneuver	1	1	4	1	0	0	8
11. Animal/no maneuver	21	36	39	53	216	5	369
12. Pedestrian/maneuver	3	1	3	10	6	0	23
13. Pedestrian/no maneuver	9	10	22	63	798	11	913
14. Cyclist/maneuver	0	3	4	3	0	0	10
15. Cyclist/no maneuver	1	1	5	35	53	$1	$96
16. Backing into vehicle	556	316	146	117	202	25	1,362
17. Turning/same direction	747	971	753	1,026	2,202	102	5,800
18. Parking/same direction	148	157	110	173	349	10	948
19. Changing lanes/same direction	1,225	1,205	816	1,180	1,957	62	6,446
20. Drifting/same lane	426	470	410	499	1,314	29	3,148
21. Opposite direction/maneuver	19	65	100	299	1,794	9	2,286
22. Opposite direction/no maneuver	263	619	1,126	2,473	17,403	84	21,967
23. Rear-end/striking maneuver	288	530	263	354	537	52	2,023
24. Rear-end/LVA	73	183	84	161	180	15	697
25. Rear-end/LVM	640	1,692	1,117	1,405	4,060	161	9,076
26. Rear-end/LVD	1,297	3,230	1,550	2,026	1,503	221	9,827

Pre-Crash Scenario	Cost by Injury Type (millions of dollars)						
	No Injury	Possible Injury	Non-Incapacitating	Incapacitating	Fatal	Injured, Unknown Severity	Total
27. Rear-end/LVS	3,028	8,529	4,005	4,765	3,177	446	23,951
28. LTAP/OD at signal	518	1,960	2,404	2,787	2,360	222	10,252
29. Turn right at signal	121	129	92	75	556	19	991
30. Left turn across path/ opposite direction at non-signal	503	1,672	1,861	2,618	5,052	134	11,840
31. SCP at non-signal	1,991	5,044	4,934	6,872	13,504	430	32,776
32. Turn at non-signal	153	151	122	95	177	3	701
33. Evasive maneuver/maneuver	30	76	37	44	93	9	290
34. Evasive maneuver/no maneuver	93	179	235	220	829	13	1,569
35. Rollover	0	3	13	22	369	1	410
36. Non-collision no impact	62	41	43	69	55	1	273
37. Object contacted/maneuver	2	1	6	6	17	0	32
38. Object contacted/no maneuver	36	36	32	43	239	1	386
39. Hit and run	14	8	11	20	18	0	71
40. Other—rear-end	6	3	8	2	0	0	19
41. Other—sideswipe	8	4	0	6	16	0	35
43. Other—turn across path	4	0	0	0	0	0	4
44. Other—turn into path	5	7	5	3	0	1	21
46. Other	89	74	67	86	211	17	544
Total	13,407	31,233	24,446	33,266	70,252	2,508	175,111

Note: Pre-crash scenarios for which there were no annual observed crashes were omitted from this table.

Table 26. Single-vehicle pre-crash scenarios by vehicle type— annual crashes (observed)

Single-Vehicle Pre-Crash Scenario	Light Vehicles	Transit Vehicles	Special Vehicles	Single-Unit Trucks	Combined-Unit Trucks	Other Vehicles	Total
1. No driver present	6	0	0	1	0	1	7
2. Vehicle failure	364	1	1	33	39	30	467
3. Control loss/vehicle action	690	1	1	21	41	141	893
4. Control loss/no vehicle action	4,803	4	4	87	200	522	5,620
5. Running red light	16	0	0	1	1	0	18
6. Running stop sign	39	0	0	0	1	1	41
7. Road edge departure/maneuver	509	5	4	34	51	54	656
8. Road edge departure/no maneuver	4,141	8	8	112	134	291	4,693
9. Road edge departure/backing	339	2	4	30	8	22	404
10. Animal/maneuver	87	0	0	7	1	7	101
11. Animal/no maneuver	1,548	2	2	11	30	94	1,686
12. Pedestrian/maneuver	505	10	2	16	4	25	562
13. Pedestrian/no maneuver	1,109	12	4	24	10	68	1,226
14. Cyclist/maneuver	534	2	2	11	3	14	564
15. Cyclist/no maneuver	658	8	2	10	4	36	716
16. Backing into vehicle	3	0	0	1	0	0	3
18. Parking/same direction	2	0	0	0	0	2	4
19. Changing lanes/same direction	299	0	0	12	25	57	392
21. Opposite direction/maneuver	14	0	0	1	0	4	18
22. Opposite direction/no maneuver	184	0	0	12	8	44	248
23. Rear-end/striking maneuver	18	0	0	2	2	5	26

Single-Vehicle Pre-Crash Scenario	Light Vehicles	Transit Vehicles	Special Vehicles	Single-Unit Trucks	Combined-Unit Trucks	Other Vehicles	Total
25. Rear-end/lead vehicle constant speed	25	0	0	2	2	5	34
26. Rear-end/LVD	60	0	0	10	10	39	119
27. Rear-end/LVS	61	0	0	5	9	23	97
Straight crossing path at non-signal	23	0	0	0	1	14	39
Turn at non-signal	54	1	0	3	1	39	97
33. Evasive maneuver/maneuver	30	0	0	2	1	6	38
34. Evasive maneuver/no maneuver	165	0	0	11	11	31	217
35. Rollover	41	1	0	11	23	74	149
36. Non-collision no impact	126	2	1	6	36	20	191
37. Object contacted/ maneuver	185	1	4	71	8	57	326
38. Object contacted/no maneuver	436	1	1	57	69	52	615
Total	17,069	59	38	598	730	1,771	20,265

Note: Pre-crash scenarios for which there were no annual observed crashes were omitted from this table.

Table 27. Single-vehicle pre-crash scenarios by vehicle type—annual national crashes (weighted)

Single-Vehicle Pre-Crash Scenario	Light Vehicles	Transit Vehicles	Special Vehicles	Single-Unit Trucks	Combined-Unit Trucks	Other Vehicles	Total
1. No driver present	540	3	0	26	1	32	602
2. Vehicle failure	34,928	18	10	896	1,452	1,588	38,894
3. Control loss/vehicle action	69,501	75	153	589	1,116	5,739	77,173
4. Control loss/no vehicle action	427,253	363	168	2,863	5,335	21,732	457,714
5. Running red light	860	0	0	4	4	1	868
6. Running stop sign	3,459	0	0	0	46	81	3,586
7. Road edge departure/ maneuver	64,416	733	944	5,876	7,932	5,241	85,141
8. Road edge departure/no maneuver	375,558	1,386	1,588	11,765	6,545	20,234	417,076
9. Road edge departure/backing	71,128	471	830	6,667	1,219	5,225	85,540
10. Animal/maneuver	15,049	0	0	1,705	210	880	17,844
11. Animal/no maneuver	283,464	392	405	1,526	3,272	6,004	295,063
12. Pedestrian/maneuver	18,608	263	72	494	47	1,145	20,629
13. Pedestrian/no maneuver	37,551	326	57	329	193	2,145	40,603
14. Cyclist/maneuver	20,255	88	41	140	29	470	21,022
15. Cyclist/no maneuver	27,160	131	140	116	97	1,112	28,757
16. Backing into vehicle	354	0	0	5	0	12	371
18. Parking/same direction	129	0	0	1	0	33	163
19. Changing lanes/same direction	28,174	0	0	318	580	1,679	30,751
21. Opposite direction/maneuver	1,567	0	0	74	0	279	1,920
22. Opposite direction/no maneuver	21,944	26	0	910	724	1,988	25,593
23. Rear-end/striking maneuver	2,233	0	0	118	105	280	2,736
25. Rear-end/lead vehicle constant speed	1,913	0	0	48	58	112	2,131
26. Rear-end/LVD	6,020	4	0	221	241	1,286	7,771
27. Rear-end/LVS	6,770	0	1	122	280	885	8,057
31. Straight crossing path at non-signal	3,131	5	55	0	97	484	3,773

Single-Vehicle Pre-Crash Scenario	Light Vehicles	Transit Vehicles	Special Vehicles	Single-Unit Trucks	Combined-Unit Trucks	Other Vehicles	Total
32. Turn at non-signal	5,795	7	4	221	134	1,410	7,572
33. Evasive maneuver/maneuver	3,005	0	4	83	137	395	3,624
34. Evasive maneuver/ no maneuver	16,274	9	5	480	377	1,219	18,363
35. Rollover	2,684	6	5	159	624	2,648	6,126
36. Non-collision no impact	16,259	51	9	355	3,256	1,242	21,171
37. Object contacted/maneuver	34,235	190	1,085	17,964	1,218	13,527	68,220
38. Object contacted/ no maneuver	57,493	136	64	8,643	4,610	7,863	78,808
Total	1,657,709	4,684	5,641	62,719	39,938	106,972	1,877,663

Note: Pre-crash scenarios for which there were no annual observed crashes were omitted from this table.

Table 28. Multi-vehicle pre-crash scenarios by vehicle type—annual crashes (observed)

Multi-Vehicle Pre-Crash Scenario	Light Vehicles	Transit Vehicles	Special Vehicles	Single-Unit Trucks	Combined-Unit Trucks	Other Vehicles	Not Reported	Total
2. Vehicle failure	227	1	1	27	56	5	52	367
3. Control loss/vehicle action	524	1	1	25	78	14	56	700
4. Control loss/no vehicle action	1,433	3	6	68	204	40	215	1,967
5. Running red light	5,348	26	9	209	158	113	1	5,864
6. Running stop sign	886	4	3	27	18	18	0	956
7. Road edge departure/ maneuver	27	0	0	2	2	1	30	60
8. Road edge departure/ no maneuver	81	0	0	3	6	2	90	181
10. Animal/maneuver	4	0	0	0	0	2	1	7
11. Animal/no maneuver	74	0	0	3	3	4	20	104
12. Pedestrian/maneuver	8	0	0	1	1	1	4	14
13. Pedestrian/no maneuver	49	0	0	2	2	4	23	80
14. Cyclist/maneuver	4	0	0	0	0	0	2	6
15. Cyclist/no maneuver	10	0	0	0	0	0	7	17
16. Backing into vehicle	1,115	5	8	118	71	31	11	1,358

Table 28. (Continued)

Multi-Vehicle Pre-Crash Scenario	Light Vehicles	Transit Vehicles	Special Vehicles	Single-Unit Trucks	Combined-Unit Trucks	Other Vehicles	Not Reported	Total
17. Turning/same direction	2,449	21	11	198	317	128	0	3,124
18. Parking/same direction	433	4	2	25	28	18	2	511
19. Changing lanes/same direction	4,079	28	10	363	848	166	7	5,501
20. Drifting/same lane	1,514	16	4	148	331	73	24	2,109
21. Opposite direction/ maneuver	206	1	2	8	9	9	2	237
22. Opposite direction/ no maneuver	2,149	6	13	108	110	57	13	2,455
23. Rear-end/striking maneuver	987	3	3	51	65	37	9	1,155
24. Rear-end/LVA	266	1	3	12	16	6	0	302
25. Rear-end/LVM	3,073	7	8	189	333	104	8	3,721
26. Rear-end/LVD	5,360	15	10	269	351	125	25	6,155
27. Rear-end/LVS	12,512	45	32	442	392	183	99	13,706
28. Left turn across path/opposite direction at signal	4,423	12	4	80	94	105	0	4,718
29. Turn right at signal	384	4	1	25	28	14	0	456
30. Left turn across path/opposite direction at non-signal	3,537	8	6	68	44	145	0	3,807
31. Straight crossing path at non-signal	10,586	31	32	290	235	365	28	11,566
32. Turn at non-signal	415	3	1	22	24	17	3	485
33. Evasive maneuver/maneuver	121	0	1	4	7	3	7	142
34. Evasive maneuver/ no maneuver	431	1	1	19	30	38	62	581
35. Rollover	3	0	0	1	2	8	13	27
36. Non-collision no impact	58	0	0	24	46	5	133	266
37. Object contact/maneuver	8	0	0	1	1	0	9	19
38. Object contact/no maneuver	93	0	0	13	28	6	59	199
39. Hit and run	37	1	0	5	3	5	2	53
40. Other—rear-end	20	0	0	0	1	0	4	26
41. Other—sideswipe	20	0	0	3	3	1	0	26
43. Other—turn across path	7	0	0	1	2	0	0	10
44. Other—turn into path	17	0	0	0	1	0	0	18
Total	62,971	246	170	2,850	3,944	1,847	1,020	73,047

Note: Pre-crash scenarios for which there were no annual observed crashes were omitted from this table.

Table 29. Multi-vehicle pre-crash scenarios by vehicle type—annual national crashes (weighted)

Multi-Vehicle Pre-Crash Scenario	Light Vehicles	Transit Vehicles	Special Vehicles	Single-Unit Trucks	Combined-Unit Trucks	Other Vehicles	Not Reported	Total
2. Vehicle failure	18,108	89	115	549	1,866	244	2,791	23,762
3. Control loss/ vehicle action	55,741	92	125	884	2,153	1,005	2,663	62,664
4. Control loss/no vehicle action	106,048	227	638	2,689	4,518	2,812	9,228	126,160
5. Running red light	447,840	2,307	591	4,891	4,029	8,307	62	468,027
6. Running stop sign	78,673	236	388	828	669	879	4	81,676
7. Road edge departure/ maneuver	1,183	0	0	199	74	22	1,479	2,958
8. Road edge departure/no maneuver	3,998	0	0	85	141	76	4,299	8,599
10 Animal/maneuver	456	5	0	0	0	177	169	807
11 Animal/ no maneuver	9,140	0	0	113	149	193	2,534	12,129
12 Pedestrian/ maneuver	819	71	0	69	71	66	194	1,291
13 Pedestrian/ no maneuver	4,976	0	0	9	18	189	530	5,723
14 Cyclist/maneuver	253	0	0	0	0	0	57	311
15. Cyclist/no maneuver	748	0	0	65	0	3	184	1,000
16. Backing into vehicle	232,283	1,103	1,757	11,781	7,568	4,605	2,633	261,731
17. Turning/same direction	354,003	3,710	1,857	12,116	16,762	10,172	1	398,621
18. Parking/same direction	69,973	808	236	1,898	1,217	1,865	179	76,176
19. Changing lanes/same direction	550,798	4,586	1,839	21,185	27,774	13,059	625	619,866
20. Drifting/same lane	187,281	3,023	772	10,031	11,285	6,653	2,872	221,918
21. Opposite direction/maneuver	15,015	27	120	278	248	799	281	16,768
22. Opposite direction/ no maneuver	172,669	351	1,722	6,975	6,749	4,930	1,733	195,129
23. Rear-end/striking maneuver	149,967	471	370	2,607	1,558	4,165	1,036	160,174

Table 29. (Continued)

Multi-Vehicle Pre-Crash Scenario	Light Vehicles	Transit Vehicles	Special Vehicles	Single-Unit Trucks	Combined-Unit Trucks	Other Vehicles	Not Reported	Total
24. Rear-end/LVA	41,766	25	311	741	561	695	0	44,098
25. Rear-end/LVM	375,424	347	798	6,486	8,081	7,856	438	399,429
26. Rear-end/LVD	742,967	1,254	1,127	9,743	6,754	10,136	979	772,960
27. Rear-end/LVS	1,798,898	5,082	3,793	18,028	11,281	17,326	5,682	1,860,090
28. Left turn across path/opposite direction at signal	388,802	942	205	2,316	2,612	5,546	0	400,423
29. Turn right at signal	56,905	646	279	1,877	1,019	1,233	0	61,959
30. Left turn across path/opposite direction at non-signal	351,595	945	453	3,115	2,165	6,873	0	365,148
31. Straight crossing path at non-signal	1,227,427	3,704	4,206	12,625	10,285	24,425	1,088	1,283,759
32. Turn at non-signal	70,503	547	129	2,533	1,939	1,397	529	77,577
33. Evasive maneuver/ maneuver	17,023	5	88	240	141	210	166	17,872
34. Evasive maneuver/ no maneuver	49,420	212	172	994	1,051	1,956	3,010	56,816
35. Rollover	142	0	0	14	81	214	450	901
36. Non-collision no impact	8,358	69	0	2,780	3,596	515	15,318	30,636
37. Object contact/maneuver	526	0	0	144	43	5	474	1,191
38. Object contact/no maneuver	10,188	0	77	1,897	1,683	635	3,960	18,440
39. Hit and run	4,960	139	0	867	95	588	454	7,102
40. Other—rear-end	2,741	0	4	1	1	3	283	3,032
41. Other—sideswipe	3,286	63	0	175	154	5	68	3,751
43. Other—turn across path	1,344	0	0	131	196	0	0	1,672
44. Other—turn into path	2,743	0	0	0	74	4	13	2,834
Total	7,614,990	31,085	22,170	141,957	138,662	139,844	66,469	8,155,177

Note: Pre-crash scenarios for which there were no annual observed crashes were omitted from this table.

This report identifies the magnitude, characteristics, and cost of crashes that would be targeted with currently proposed V2I safety application areas. It also identifies the magnitude, characteristics, and cost of the remaining crashes that are not targeted by a currently proposed V2I for safety application area (unaddressed crashes) to help identify potential new applications or modifications to proposed applications. The results indicate that the currently identified V2I application areas are well conceived and can potentially reduce crashes and crash costs. However, many crashes are not addressed in this report, and characteristics of unaddressed crashes are presented in this report provide a starting point for identifying either new applications or modifications to current applications.

REFERENCES

[1] National Highway Traffic Safety Administration. (2009). *National Automotive Sampling System/General Estimates System (NASS GES)*, Washington, DC. Obtained from: ftp://ftp.nhtsa.dot.gov/GES/.

[2] Federal Highway Administration. (2011). *Highway Safety Information System (HSIS)*, Washington, DC. Obtained from: http://www.hsisinfo.org.

[3] National Highway Traffic Safety Administration. (2009). *Fatality Analysis Reporting System (FARS)*, Washington, DC. Obtained from: ftp://ftp.nhtsa.dot.gov/FARS/.

[4] Shelton, T.S. (1993). *Imputation in the NASS General Estimates System*, Report No. DOT HS 807 985, U.S. Department of Transportation, Washington, DC.

[5] National Highway Traffic Safety Administration. (2009). *National Automotive Sampling System (NASS) General Estimates System (GES) Analytical User's Manual 1988–2008*, U.S. Department of Transportation, Washington, DC.

[6] Shelton, T.S. (1991). *National Accident Sampling System General Estimates System Technical Note*, Report No. DOT HS 807 796, U.S. Department of Transportation, Washington, DC.

[7] Najm, W.G., Smith, J.D., and Yanagisawa, M. (2007). *Pre-Crash Scenario Typology for Crash Avoidance Research*, Report No. DOT HS 810 767, U.S. Department of Transportation, Washington DC.

[8] DaSilva, M.P., Smith, J.D., and Najm, W.G. *Heavy Vehicle Pre-Crash Scenario Typology for Crash Avoidance Research*, U.S. Department of

Transportation, Washington DC. (Manuscript submitted in 2009 for publication and in-press).

[9] National Highway Traffic Safety Administration. (2008). *Preliminary Regulatory Evaluation: FMVSS No. 128 Motorcycle Helmet Labeling*, U.S. Department of Transportation, Washington, DC.

[10] North America Operations Crash Avoidance Department. (1997). *44 Crashes*, Version 3.0, General Motors Corporation, Warren, MI.

[11] Najm, W.G., Sen, B., Smith, J.D., and Campbell, B.N. (2003). *Analysis of Light Vehicle Crashes and Pre-Crash Scenarios Based on the 2000 General Estimates System*, Report No. DOT HS 809 573, U.S Department of Transportation, Washington, DC.

[12] Najm, W.G., Koopmann, J., Smith, J.D., and Brewer, J. (2010). *Frequency of Target Crashes for IntelliDrive Safety Systems*, Report No. DOT HS 811 381, U.S. Department of Transportation, Washington, DC.

[13] Chang, J., Cohen, D., Blincoe, L., Subramanian, R., and Lombardo, L. (2007). *CICAS-V Research on Comprehensive Costs of Intersection Crashes*, Report No. 07-0016, U.S. Department of Transportation, Washington, DC.

[14] National Highway Traffic Safety Administration. (2009). *National Automotive Sampling System Crashworthiness Data System (NASS CDS)*, Washington, DC. Obtained from http://www.nhtsa.dot.gov/people/ncsa/ nass_cds.html.

[15] Federal Highway Administration. (2003). *Intersection Collision Avoidance Study Final Report*, Report No. FHWA/TX-03/4027-1, U.S Department of Transportation, Washington, DC.

[16] Mitretek Systems Analysis. (2005). *Cooperative Intersection Collision Avoidance System: Distribution of Crashes by Intersections—An Exploratory Analysis*, Draft Version 1.0, Falls Church, VA.

[17] NCUTLO. (2000). *Uniform Vehicle Code*, Millennium Ed., National Committee on Uniform Traffic Laws and Ordinances, Alexandria, VA.

INDEX

D

E

F

G

H

I

J

L

M

N

W